Diagnostic Biomedical Optics

Fundamentals and applications

Online at: https://doi.org/10.1088/978-0-7503-2364-2

IOP Series in Advances in Optics, Photonics and Optoelectronics

SERIES EDITOR

Professor Rajpal S Sirohi Consultant Scientist

About the Editor

Rajpal S Sirohi is currently working as a faculty member in the Department of Physics, Alabama A&M University, Huntsville, AL, USA. Prior to this, he was a consultant scientist at the Indian Institute of Science, Bangalore, and before that he was Chair Professor in the Department of Physics, Tezpur University, Assam. During 2000–2011, he was an academic administrator, being vice-chancellor to a couple of universities and the director of the Indian Institute of Technology, Delhi. He is the recipient of many international and national awards and the author of more than 400 papers. Dr Sirohi is involved with research concerning optical metrology, optical instrumentation, holography, and the speckle phenomena.

About the series

Optics, photonics, and optoelectronics are enabling technologies in many branches of science, engineering, medicine, and agriculture. These technologies have reshaped our outlook and our ways of interacting with each other, and have brought people closer together. They help us to understand many phenomena better and provide deeper insight into the functioning of nature. Further, these technologies themselves are evolving at a rapid rate. Their applications encompass very large spatial scales, from nanometres to the astronomical scale, and a very large temporal range, from picoseconds to billions of years. This series on advances in optics, photonics, and optoelectronics aims to cover topics that are of interest to both academia and industry. Some of the topics to be covered by the books in this series include biophotonics and medical imaging, devices, electromagnetics, fibre optics, information storage, instrumentation, light sources, charge-coupled devices (CCDs) and complementary metal oxide semiconductor (CMOS) imagers, metamaterials, optical metrology, optical networks, photovoltaics, free-form optics and its evaluation, singular optics, cryptography, and sensors.

About IOP ebooks

The authors are encouraged to take advantage of the features made possible by electronic publication to enhance the reader experience through the use of color, animation, and video and by incorporating supplementary files in their work.

A full list of titles published in this series can be found here: https://iopscience.iop.org/bookListInfo/series-on-advances-in-optics-photonics-and-optoelectronics.

Diagnostic Biomedical Optics

Fundamentals and applications

Edited by

Murukeshan Vadakke Matham

Centre for Optical and Laser Engineering, School of Mechanical and Aerospace Engineering, Nanyang Technological University, 50 Nanyang Avenue, Singapore 639798, Singapore

C S Suchand Sandeep

Department of Physics, Manipal Institute of Technology, Manipal Academy of Higher Education, Manipal, Udupi, Karnataka State, 576104, India

Maria Merin Antony

Centre for Optical and Laser Engineering, School of Mechanical and Aerospace Engineering, Nanyang Technological University, 50 Nanyang Avenue, Singapore 639798, Singapore

Manojit Pramanik

Department of Electrical and Computer Engineering, Iowa State University, 2520 Osborn Drive, Ames, IA, USA

Santhosh Chidangil

Centre of Excellence for Biophotonics, Manipal Institute of Applied Physics, Manipal Academy of Higher Education, Manipal, Karnataka 576104, India

IOP Publishing, Bristol, UK

ISBN 978-0-7503-2364-2 (ebook)
ISBN 978-0-7503-2362-8 (print)
ISBN 978-0-7503-2365-9 (myPrint)
ISBN 978-0-7503-2363-5 (mobi)

DOI 10.1088/978-0-7503-2364-2

Version: 20250701

IOP ebooks

British Library Cataloguing-in-Publication Data: A catalogue record for this book is available from the British Library.

Published by IOP Publishing, wholly owned by The Institute of Physics, London

IOP Publishing, No.2 The Distillery, Glassfields, Avon Street, Bristol, BS2 0GR, UK

US Office: IOP Publishing, Inc., 190 North Independence Mall West, Suite 601, Philadelphia, PA 19106, USA

Dedicated to all scholars and researchers whose work continues to illuminate the field of biomedical optics.

Contents

Preface

This book will cover the basics of medical imaging methods and technologies that use optics. It will also include several in-depth analyses of the relevant topics such as photoacoustics, Raman spectroscopy, and hyperspectral imaging, to name a few. Diagnostic biomedical optics is a multidisciplinary field that utilizes optical technologies to deepen our understanding of biological processes, ultimately enhancing diagnostic accuracy and improving treatment of human diseases. Conventional medical imaging methods often struggle to detect subtle tissue changes, and since each imaging modality has its own strengths and weaknesses, no single approach is suitable for all diagnostic needs. This has led to the growing demand for multimodality or hybrid imaging techniques. Advances in technology continually drive improvements in early disease diagnostics, shifting medical practices toward a 'diagnosis under a single setting' approach, rather than subjecting patients to multiple tests. The need for multimodality diagnostics has emerged to support this shift, though implementing such systems presents challenges and requires interdisciplinary collaboration.

In modern imaging, achieving high-resolution images with sufficient working distance and the ability to image around opaque obstacles is crucial. However, the integration of multiple modalities in diagnostic imaging can sometimes diminish the unique advantages of individual techniques. To address these challenges, a paradigm shift in medical diagnostics has recently occurred, focusing on enhancing various key imaging parameters. This book will explore these developments, highlighting important features and considerations in the context of bio-imaging. This book will also give clear and concise explanations of the fundamentals and the applied scientific and technological aspects of this exciting and highly significant field of diagnostic medical optics.

Acknowledgements

First, as the main editor of this book, I would like to acknowledge the support I have received from many quarters towards the completion of this book, especially from the administration of Nanyang Technological University, the school of Mechanical and Aerospace Engineering, and the Centre for Optical and Laser Engineering (COLE). In addition, I would like to acknowledge Ms Maria Merin Antony and Dr Keerthi K for going through the initial chapters/drawings. I would like to acknowledge my research students and staff who have contributed to some of the research that is included in this book, from chapters 1–4. Special mention to Ms Maria Merin Antony, Dr Mohankumar V, Dr Jeesmond, Dr Sandeep Suchand, Dr Shinoj V K, Dr Sandeep Menon P, Dr Hoong Ta Lim, Dr James Joseph, Dr Ratheesh, Dr Anant Shinde and Dr Aswin Haridas who made significant contributions to various related biomedical imaging projects that I was pursuing over the last 15 years. Contributions and support from my collaborators in various biomedical research projects are also acknowledged.

I personally want to thank my co-editors, Prof. Manojit Pramanik, Prof. Santhosh Chidangil, Prof. Suchand Sandeep, Dr Maria Merin Antony, and co-authors of their respective book chapters, for their contribution with such a short notice.

Acknowledgements will not be completed without mentioning the tremendous financial support I have received through various funding agencies for the various biomedical research projects. They include Ministry of Education (MOE, Singapore), National Research Foundation, National Medical Research Council (NMRC), A*STAR Singapore, and some of the leading industries.

Finally, I would like to thank my wife and my children for their patience, support and encouragement to complete this edited book. The co-editors of this book would also like to thank their families and research groups for their patience and support.

Your suggestions, corrections and contributions will be appreciated and reflected in the later editions of this book.

Dr Murukeshan Vadakke Matham
(FInstPhy, FSPIE, FOptica, FOSI)

Editor biographies

Professor Murukeshan Vadakke Matham

Professor Murukeshan Vadakke Matham, the main editor of this book, serves as the Director of the Center for Optical and Laser Engineering (COLE), Deputy Director of The Photonics Institute (TPI), and Associate Professor of School of Mechanical and Aerospace Engineering, NTU Singapore. He leads a research group specializing in biomedical optics, nanoscale optics, laser-based micro and nanoscale fabrication, and applied optics for metrology. Professor Murukeshan has an extensive academic profile with over 28 years of professional experience, including three years of postdoctoral work and 25 years of teaching. He has published over 215 international journal papers and over 200 international conference proceedings/presentations, co-authored two books, and filed or awarded 26 patents and 16 innovation disclosures. To date, he has supervised 28 PhD students and 30 Masters (MEng by research and Master of Science in Engineering) students. He is also Associate Editor for Optical Engineering and the International Journal of Optomechatronics (IJO). He serves in international committees related to optics at different capacities. Professor Murukeshan has received more than 25 prestigious international recognitions or awards. He is a Fellow of SPIE, Fellow of OPTICA (USA), a Fellow of the Institute of Physics (UK), and a Distinguished Fellow of the Optical Society of India.

Professor Suchand Sandeep

Professor Suchand Sandeep is an Associate Professor at the Department of Physics at Manipal Institute of Technology, Manipal Academy of Higher Education, India. Dr Sandeep received his Master's degree in physics from the Indian Institute of Technology Madras (IIT Madras) and PhD degree in physics from the Raman Research Institute, Bangalore, India. He has over 10 years of postdoctoral research experience in renowned institutions including Delft University of Technology (TU Delft, the Netherlands), University of Potsdam (Uni Potsdam, Germany), and Nanyang Technological University (NTU, Singapore). He is a recipient of various international fellowships, including two Marie Curie individual fellowships from the European Union and a Young Researcher Fellowship from the Ministry of Education, Italy. He has co-authored over 75 international journal publications, 20 conference proceedings, and is a co-inventor in six patent applications and has submitted seven technology disclosures. He is currently a guest editor of Polymers for Advanced Technologies. His areas of expertise are photonics, renewable energy, biomedical optics, additive manufacturing, and ultrafast phenomena.

Dr Maria Merin Antony

Dr Maria Merin Antony is a Postdoctoral Research Fellow at Center for Optical and Laser Engineering (COLE), School of Mechanical and Aerospace Engineering, NTU Singapore. She was awarded her PhD degree in optical and laser engineering from Nanyang Technological University, Singapore in 2025. She received B.Tech. and M.Tech. degrees in electrical and electronics engineering from the Cochin University of Science and Technology, Kochi, India, in 2015 and 2019, respectively. She was an Embedded System Engineer with Robert Bosch, Coimbatore, India, from 2015 to 2017. Her current research interests include hyperspectral imaging (HSI), nonlinear optics, and high-resolution optical imaging. She has co-authored over 18 international publications (consisting of journal publications, conference proceedings and book chapter) and is a co-inventor in two patent applications and had submitted three technology disclosures. Her areas of expertise are hyperspectral imaging, high-resolution imaging, nonlinear optics and agriphotonics. She is also involved as Executive Committee member at Society of Women Engineers, NTU and the Agriphotonics Technical Group, Optica. She has been awarded the WiEST development grant focusing on development of the early career women scientists and engineers.

Professor Manojit Pramanik

Professor Manojit Pramanik joined the Department of Electrical and Computer Engineering (ECpE) at Iowa State University from spring 2023 and currently holding the position of Northrop Grumman Associate Professor. He also has an appointment as a Faculty of Biomedical Engineering (BME) at Iowa State University. Prior to that, he was at Nanyang Technological University (NTU), Singapore from 2014 to 2022. He also served as Assistant Professor in the Department of Electrical Engineering at Indian Institute of Science (IISc), Bangalore, India 2012–13. His industry experiences include two years at General Electric Global Research (GRC), Bangalore, India and one year at Philips Medical System, Bangalore, India. His research interest is in medical imaging systems, photoacoustic and thermoacoustic imaging, image reconstruction, machine learning, medical image processing, contrast agents, molecular imaging, Monte Carlo simulation for light–tissue interaction, biomedical optics, biomedical device design, clinical application areas such as breast cancer imaging, brain imaging, pancreatic cancer, diabetes, treatment monitoring, etc. He has more than 250 international journal and conference publications and presentations. He serves as an Associate Editor of the *Journal of Biomedical Optics*, *Biomedical Optics Express*, and as an Editorial Board Member of *Photoacoustics*. He also served as Guest Editor for special issues in the *Journal of Biomedical Optics*, *Biomedical Engineering Letters*, *IEEE OJIM*. He is a Fellow of SPIE, Senior Member of IEEE, and life member of Optica. Professor Pramanik is also the inaugural 'Biodesign Faculty Fellow' offered by the Singapore Biodesign Programme (in collaboration with Stanford Biodesign Fellowship).

Professor Santhosh Chidangil

Professor Santhosh Chidangil is Professor of Physics and Coordinator of the Centre of Excellence for Biophotonics at Manipal Academy of Higher Education (MAHE), India. His research is centered on biomedical instrumentation, with a focus on optical pathology techniques and liquid biopsy approaches for the early detection of cancer, as well as diagnostics for ophthalmological and cardiovascular diseases. With over three decades of academic and research leadership, he has successfully led more than 15 funded projects and published over 230 scientific papers. Dr Chidangil has guided 14 PhD scholars, over 15 MTech students, trained more than 20 Indian Academy Summer Research Fellows, and supervised over 20 IAESTE international research interns. He was instrumental in establishing a world-class Biophotonics Research Centre at MAHE. He served as Sectional President of the Indian Science Congress in 2020 and as India Chair for the BRICS Workshop on Biophotonics held during 2021–2024. Dr Chidangil is a member of several prestigious scientific societies, including OSA, SPIE, the Indian Laser Association, the Indian Physics Association, ISRAPS, and the Indian Nuclear Society.

List of contributors

Santhosh Chidangil
Professor, Centre of Excellence for Biophotonics, Manipal Institute of Applied Physics, Manipal Academy of Higher Education, Manipal, Karnataka 576104, India

Avishek Das
Department of Electrical and Computer Engineering, Iowa State University, 2520 Osborn Drive, Ames, IA, United States of America

Katherine Gisi
Department of Electrical and Computer Engineering, Iowa State University, 2520 Osborn Drive, Ames, IA, United States of America

Murukeshan Vadakke Matham
Director, Centre for Optical and Laser Engineering, School of Mechanical and Aerospace Engineering, Nanyang Technological University, 50 Nanyang Avenue, Singapore 639798, Singapore

Maria Merin Antony
Centre for Optical and Laser Engineering, School of Mechanical and Aerospace Engineering, Nanyang Technological University, 50 Nanyang Avenue, Singapore 639798, Singapore

N Mithun
Centre of Excellence for Biophotonics, Manipal Institute of Applied Physics, Manipal Academy of Higher Education, Manipal, Karnataka 576104, India

Manojit Pramanik
Associate Professor, Department of Electrical and Computer Engineering, Iowa State University, 2520 Osborn Drive, Ames, IA, United States of America

K P Sreejith
Centre of Excellence for Biophotonics, Manipal Institute of Applied Physics, Manipal Academy of Higher Education, Manipal, Karnataka 576104, India

C S Suchand Sandeep
Associate Professor, Department of Physics, Manipal Institute of Technology, Manipal Academy of Higher Education, Manipal, Udupi, Karnataka State, 576104, India

Vijitha Periyasamy
Department of Electrical and Computer Engineering, Iowa State University, 2520 Osborn Drive, Ames, IA, United States of America

List of abbreviations

Abbreviations	Expansion
1-D/1D	One-dimensional
2-D/2D	Two-dimensional
2PA	Two photon absorption
2PEM	Two photon excitation microscopy
3-D/3D	Three-dimensional
AI	Artificial intelligence
ALK	Anterior lamellar keratoplasty
AS	Anterior segment
ASTM	American Society for Testing and Materials
CCD	Charge-coupled device
CT	Computed tomography
DMD	Digital micro-mirror device
DoF	Depth of field
DOT/DOI	Diffused optical tomography/imaging
DRS	Diffuse reflectance spectroscopy
EMCCD	Electron multiplying charge-coupled device
ES-SIM	Embedded speckle structured illumination microscopy
FD	Frequency domain
FT	Fourier transform
FFT	Fast Fourier transform
FOIB	Fiber optic imaging bundle
FOT	Fluorescence optical tomography
FMT	Fluorescence molecular tomography
FWHM	Full-width half maximum
MRI	Magnetic resonance imaging
FPS	Frames per second
GRIN	Gradient index
GI	Gastrointestinal
GUI	Graphic user interface
H&E	Hematoxylin and eosin
HS	Hyperspectral
HSI	Hyperspectral imaging
ICA	Independent component analysis
ICE	Iridocorneal endothelial
IFS	Intrinsic fluorescence spectroscopy
LED	Light-emitting diode
LSFM	Light sheet fluorescence microscopy
LSR	Laser speckle reducer
LSS	Light scattering spectroscopy
MC	Monte Carlo

(*Continued*)

(*Continued*)

Abbreviations	Expansion
MPM	Multiphoton microscopy
MRI	Magnetic resonance imaging
NA	Numerical aperture
ND	Normalised difference
NIR	Near infrared
NP	Nanoparticles
NMR	Nuclear magnetic resonance
OCT	Optical coherence tomography
OS-SIM	Optical sectioning-structured illumination microscope
OTF	Optical transfer function
PAI	Photoacoustic imaging
PALM	Photoactivation localization microscopy
PCA	Principal component analysis
PCR	Polymerase chain reaction
PDMS	Polydimethylsiloxane
PET	Positron emission tomography
PK	Penetrating keratoplasty
PLSR	Partial least square regression
PPD	Posterior polymorphous dystrophy
PSF	Point spread function
RI	Refractive index
RAM	Random access memory
RMS	Root mean square
SAM	Spectral angle mapper
SIM	Structured illumination microscopy
SLM	Spatial light modulator
SPECT	Single photon emission computed tomography
SNR	Signal-to-noise ratio
SSD	Solid state drive
STED	Stimulated emission depletion
STORM	Stochastic optical reconstruction microscopy
SVM	Support vector machine
SWIR	Short wave infrared
TD	Time domain
TEM	Transmission electron microscopy
USAF	United States Air Force
UV	Ultraviolet
VIS	Visible
WD	Working distance
WHO	World Health Organisation
XCT	X-ray coherence tomography

IOP Publishing

Diagnostic Biomedical Optics

Fundamentals and applications

Murukeshan Vadakke Matham, C S Suchand Sandeep, Maria Merin Antony, Manojit Pramanik and Santhosh Chidangil

Chapter 1

Introduction

Murukeshan Vadakke Matham

This introductory chapter presents an overview of medical imaging technologies using optics, with detailed discussions on key topics. Diagnostic biomedical optics is an interdisciplinary field leveraging optical methods to enhance our understanding of biological processes, with the ultimate goal of improving diagnostic accuracy and treatment outcomes. Traditional imaging approaches often fall short in detecting subtle tissue changes, and no single modality addresses all diagnostic needs, driving the demand for multimodality or hybrid imaging techniques. This chapter examines recent innovations and emphasizes the importance of key imaging parameters, briefly covering both ionization- and non-ionization-based imaging modalities, setting the stage for further exploration of the science and technology behind diagnostic medical optics with a special emphasis on optical imaging modalities.

1.1 Background

Light–tissue interactions and diagnostics play a crucial role in health and disease management by aiding in early disease detection and prevention, assessment of the injury, interventions, guided surgeries, and post-treatment monitoring invasively. Over the years, numerous tissue diagnostic procedures have evolved and are widely used in clinical investigations, contributing significantly to innovations in biomedical research. These methods and procedures can be performed *ex vivo*, *in vivo* or *in vitro*, so as to explore various tissue properties along with their physical, mechanical and thermal attributes. Though these diagnostic approaches can reveal a wealth of information about tissue characteristics, they often fail to accurately localize targeted attributes. This limitation has led to a change from traditional assay-based methods to techniques that are based on imaging (Ntziachristos *et al* 2005). Imaging technologies offer powerful tools for visualizing complex and dynamic biological processes, thereby accelerating both research and clinical

doi:10.1088/978-0-7503-2364-2ch1

diagnosis. By revealing the structural and molecular behaviour of tissue components, these techniques are invaluable for early disease detection.

A wide range of imaging modalities were developed in the recent past to support tissue diagnostics, with each modality relying on specific types of energy input to the tissue. Although many medical imaging technologies are available or reported, they often struggle to provide satisfactory results for early disease diagnosis due to the inherent structural and molecular complexity of tissues. Tissues consist of diverse and intricate structures and molecules at various depths, leading to a range of biophysical and biochemical responses depending on the applied energy source. The imaging resolution and depth of penetration are dependent on wavelength and type of energy. As a result, each imaging modality is designed to operate within defined parameters and is suited for specific applications. For instance, x-rays are ideal for imaging hard tissues throughout the body, magnetic resonance imaging (MRI) is best suited for soft tissue imaging, ultrasound excels in detecting acoustic variations in tissue, and optical imaging offers the highest resolution but suffers from poor penetration depth (Ntziachristos *et al* 2002, Judenhofer *et al* 2008, Ng *et al* 2012). Therefore, no single imaging modality can provide the required comprehensive structural and molecular information across varying spatial resolutions and tissue depths. In order to address this, by sequentially using multiple imaging modalities, we can enable image registration techniques to co-register data from different sources (Hill *et al* 2001). This necessity led to the development of multimodal imaging systems. These multimodal imaging systems combine multiple imaging modalities such as positron emission tomography (PET), computed tomography (CT), single-photon emission computed tomography (SPECT), fluorescence molecular tomography (FMT), x-ray computed tomography (XCT), magnetic resonance imaging (MRI), and fluorescence optical tomography (FOT) in a single setup. Such systems enable near-simultaneous imaging, providing multi-level and multi-scale information about tissue structure and function.

In recent years, considerable efforts have been dedicated to developing imaging systems capable of generating multimodal images using a single instrument or detector. They offer significant advancements in combining multiple imaging modalities and include systems such as PET/CT (Beyer *et al* 2000, Townsend *et al* 2003, Nehmeh *et al* 2004), SPECT/CT (Kalki *et al* 1995, Tang *et al* 1996, MacDonald *et al* 2002, Weisenberger *et al* 2003), PET/MRI (Cherry *et al* 2008, Hofmann *et al* 2008, Judenhofer *et al* 2008, Schlemmer *et al* 2008, Ng *et al* 2012), PET/FOT (Li *et al* 2011) and FMT/XCT (Ale *et al* 2012). These systems and their details are widely reported in various research publications and other reports. Despite these advancements, there remains a lack of imaging techniques that can rely solely on non-ionizing radiation while reducing the time required for interrogation. Further, the use of ionizing radiation in these systems presents potential health risks, with long imaging times typically ranging from minutes to hours.

A promising approach involves integrating non-ionizing modalities, such as ultrasound and optical imaging, to create a system capable of delivering comprehensive diagnostic information (James *et al* 2014). Ultrasound imaging can provide structural maps of acoustic heterogeneities deep within tissue at mesoscopic (sub-millimetre) spatial resolution, while optical imaging can offer detailed structural and molecular maps at sub-micron resolution across the tissue surface.

Moreover, the penetration depth limitations associated with conventional optical imaging modalities can be overcome by employing techniques such as photoacoustic imaging. This method allows for the mapping of optical absorption heterogeneities at resolution close to ultrasound techniques throughout tissue depth and effectively reduce the gap between tissue surface and deep-tissue imaging. Together, these advancements point towards a future where non-ionizing, multimodal imaging systems offer safer, faster, and more detailed diagnostic capabilities.

1.2 Medical and bio-imaging

Recent advancements in science and technology have significantly enhanced our understanding of biological systems at the microscopic level. Currently, a wide range of diagnostic procedures are available for disease detection and monitoring. Ongoing research aims to develop methods that are minimally invasive, offering quick and accurate results that are patient friendly. In the following sections, we will briefly discuss several key diagnostic techniques that are widely used in clinical settings. These include the ionization-based techniques such as x-ray, CT, MRI, PET/SPECT, and endoscopy techniques. They play a critical role in diagnosing internal conditions and assessing disease progression within the body (Lindsley *et al* 2004, Murukeshan and Sujatha 2007, Mohankumar *et al* 2012, Lim and Murukeshan 2016).

Imaging methods using optical technologies are used for detecting changes in morphology and characteristics of the tissues and organs, as well as the ambience. There are many high-resolution microscopic techniques that are developed or reported, which include stimulated emission depletion (STED) microscopy, stochastic optical reconstruction microscopy (STORM), and confocal laser scanning microscopy (CLSM), light sheet fluorescence microscopy (LSFM), multi-photon microscope (MPM), etc. A comparative analysis of the capabilities of these microscopic techniques is tabulated in table 1.1. Commercial systems like patterned array microscopy (PAM), structured illumination microscopy (SIM) and embedded speckle structured illumination microscopy (ES-SIM) have been adopted for targeted illumination purposes, with their application remaining largely aimed at *in vitro* studies (Neil *et al* 1997, Antony *et al* 2023). The current challenge still focuses on developing techniques that can integrate high-resolution imaging concepts and technology with precise targeting capabilities suitable for broader applications, including *in vivo* studies.

With these advancements in the recent past, biomedical optics has revolutionized the way diseases are detected and diagnosed, offering non-invasive techniques to visualize tissue abnormalities. In the case of endoscopes, they are commonly used for *in vivo* imaging and come in various sizes, shapes, and flexibility. Based on their working principles, endoscopes can be broadly classified into rod lens and tip chip endoscopes, and fiberscopes. Though endoscopes have limitations in terms of reach and dimensions, they can typically achieve imaging resolutions around 0.1 mm.

Table 1.1. Various super-resolution imaging techniques.

Modality	Methodology	Resolution	Light source	Pros	Cons
SIM/ES-SIM	Structured illumination/ embedded speckle pattern illumination	100–300 nm	Laser, single wavelength	High contrast	Phototoxicity
LSFM	Thin light sheet - excitation of fluorophores within the focal volume	~100 nm	Laser, single wavelength	3D imaging of thick samples	Limited field of view
STED	Stimulated emission depletion	30–50 nm (lateral) 40–150 nm (axial)	Multiple lasers	Fast	Phototoxicity system complexity
Confocal	Rejecting out of focus light using common mode rejection using pinholes Photodamage at focal point	~200 nm	(lateral)~500 nm (axial)	Laser, single wavelength	3D imaging of thick samples
MPM	High power lasers to induce nonlinear optical processes	~60 nm	Femtosecond or picosecond Laser in IR region	Deeper sectioning than confocal microscopy	Expensive

In addition to these, several fiber optic imaging bundle (FOIB) systems have been developed for *in vivo* imaging, offering higher resolution capabilities. These FOIB-based optical techniques have also demonstrated their potential for structured illumination with potential minimally invasive or *in vivo* imaging applications. These recent scientific and technological advancements in targeted illumination and imaging methods remain primarily suitable for *in vitro* studies (Murukeshan and Sujatha 2007, Mohankumar *et al* 2012). Despite these developments, very few optical systems have been reported for controlled parametric studies involving fiber optic-based illumination of targeted cells and organelles *in vivo*. Such targeted illumination configuration and integrated imaging systems are tailored for specific bioimaging applications with potential implications.

Furthermore, a multimodality imaging approach enables the analysis of tissue conditions using multiple methods, enhancing diagnostic accuracy by cross-referencing information from different modalities. Optical-based multimodality systems offer several advantages, including reduced invasiveness, compact integration of instruments, real-time imaging and analysis, faster diagnosis, and the potential for incorporation into endoscopic setups (Murukeshan and Sujatha 2007, Lim and Murukeshan 2016).

In one study utilizing Hyperspectral Imaging (HSI) with a Liquid Crystal Tuneable Filter (LCTF), spectral mapping was performed across wavelengths from 440 to 650 nm, in 10 nm increments for each excitation wavelength (Szu *et al* 2000). Although malignant areas were distinguishable from normal tissue, the process took up to 23 s per image, making it unsuitable for real-time endoscopic diagnostics. To address this, Lindsley *et al* (2004) developed a hyperspectral endoscope using polarized light illumination, allowing the endoscopist to view either single-wavelength images or a series of fast wavelength scans in a non-contact manner. Many different spectroscopic techniques based on Intrinsic Fluorescence IFS), Diffuse Reflectance (DRS), and Light Scattering (LSS) are also developed by researchers to characterize dysplastic lesions. The disease state was determined by comparing spectral patterns with those of normal tissue. Further, attempts were made to measure the absorption and scattering properties of colon tissue *in vitro*, which can enable development of a model to predict spectral differences between normal and cancerous tissue. However, as absorption by blood plays a crucial role, it was advised to exercise caution when applying *in vitro* values to *in vivo* conditions.

Diffraction gratings and CCDs were used for spectral distribution mapping with better accuracy. However, this point-by-point scanning method is found to be time-consuming, making it less practical for real-time applications. A study proposed to employ a photodynamic diagnosis method using a halogen lamp for illumination and scanning-based imaging spectrometry, where a monochrome CCD was used to provide spectral information based on pixel position (Marra *et al* 2018). In another study, a CCD camera and monochromator were employed to detect fluorescence differences between normal and cancerous tissues, using tissue phantoms stained with fluorophores (Lagarto *et al* 2020).

Fluorescence imaging, with or without markers, has also been explored, typically using UV or short-wavelength illumination. For example, a fluorescence image of tissue was captured under UV light, and the ratio of green to red intensities in the image was used to assess cancerous conditions. Filters were used to block the illumination wavelength. In a clinical study, spectral differences between normal, adenoma, and carcinoma conditions of colon tissue were mapped using optimal excitation wavelength, and also by employing frequency domain fluorescence imaging (Dinish *et al* 2007, Weissleder and Pittet 2008, Li *et al* 2011, Ale *et al* 2012).

Another method is autofluorescence imaging, which uses specialized cameras and light sources to highlight differences between normal and diseased tissues. A notable technique involves using monochromatic CCD cameras with a xenon arc lamp, where blue light excites autofluorescence and emitted/reflected green light captures for analysis. By combining these images, normal tissue appears green while cancerous regions are highlighted in magenta. However, the accuracy of this approach depends on the severity of the abnormality and the expertise of the observer in interpreting the images. The research by Matsuo *et al* has further contributed to this field by examining the mucus gel layer on the surface of the colon. Their findings revealed that while healthy colon tissue has a thick protective layer, this gel is significantly reduced or absent in precancerous and cancerous regions (Matsuo *et al* 1997). Understanding how these structural differences affect reflectance spectroscopy could pave the way for improved endoscopic imaging techniques. By enhancing the contrast between normal and diseased tissue, these innovations could lead to more effective and accessible early detection methods for colorectal cancer.

1.3 Imaging modalities

Different imaging techniques are classified based on the type of energy, resolution or information they provide. They can be classified with respect to the applied energy, such as x-rays, acoustics, light, or positrons. In terms of their spatial resolution, they can be categorized ranging from mesoscopic to microscopic to nanoscopic levels. Furthermore, based on the type of information they provide, they can be further classified as anatomical, physiological, or molecular (Weissleder and Pittet 2008). In general, most of the medical imaging methods have relied on external electromagnetic energy, including radio waves and x-ray radiation, to interact with tissues and produce visual representations (Ale *et al* 2012, Townsend and Cherry 2001).

These imaging techniques can be grouped into ionizing and non-ionizing categories. Ionizing techniques, like x-rays and PET, use radiation that can pose health risks with prolonged exposure. In contrast, non-ionizing techniques, such as ultrasound and optical imaging, are generally safer and more suitable for repeated use in medical diagnostics. Advances in biomedical optics have significantly enhanced non-ionizing imaging, offering high-resolution, non-invasive solutions for early disease detection and clinical applications. The following section provides a brief review of the most commonly used imaging modalities, highlighting their advantages and limitations.

Imaging methods that rely on high-energy ionizing radiation generally use x-rays and gamma rays, to generate detailed images of internal tissues. Though highly effective for medical diagnostics, their use is limited by safety concerns, as repeated exposure to ionizing radiation must be carefully controlled to minimize health risks. Ionizing imaging techniques such as x-ray computed tomography (XCT) provide cross-sectional images of the body. Techniques such as nuclear imaging modalities that include PET and SPECT offer capabilities for detecting abnormalities, assessing organ function as well as guiding treatment decisions (Malko *et al* 1986, Beyer *et al* 2000, Cherry and Dahlbom 2004, Lucas *et al* 2006, Li *et al* 2011).

XCT is a diagnostic tool that employs x-ray technology for generating cross-sectional images of tissues. However, it has poor contrast when imaging soft tissues and the use of ionizing radiation poses health risks (Thomasson *et al* 2004). Sensitive PET and SPECT-based nuclear imaging modalities are capable of providing functional and molecular insights into tissues (Beyer *et al* 2000, Cherry and Dahlbom 2004, Schlemmer *et al* 2008). However, their drawbacks lie in the low spatial resolution of the order of few millimeters, longer imaging times (~hours), and the possible health hazards due to the use of radioactive isotopes (Ntziachristos *et al* 2000, Ng *et al* 2012).

Non-ionizing techniques use low-energy radiation, which will not pose significant health hazards like ionization radiation. They allow for repeated imaging without concerns about radiation exposure, and higher energy levels can be used when necessary to improve image quality and diagnostic accuracy. Most used non-ionizing imaging techniques include magnetic resonance imaging (MRI), and Ultrasound Imaging. MRI provides detailed anatomical and functional insights whereas ultrasound generally used for real-time imaging of soft tissues. From this perspective, optical imaging offers high-resolution visualization of biological structures at the cellular and molecular levels (Lu and Fei 2014, Antony *et al* 2024). MRI works based on the emission of electromagnetic (EM) waves through nuclear resonance, thereby producing detectable signals and using powerful magnets. It offers excellent soft tissue contrast. Functional MRI (fMRI) in general creates activation maps to visualize specific processes, capturing functional changes or activity. Additionally, sub-millimeter resolution between 10 and 100 μm is achievable using a small receiver coil with a stronger magnetic field.

Ultrasound imaging uses acoustic waves, providing structural imaging with good spatial resolution (of the order of few tens of μm) and shorter imaging times (seconds to minutes; Wells 2006). However, the limitations are there due to poor soft tissue contrast and limited functional and molecular imaging ability. These non-ionizing methods have become essential tools in medical diagnostics due to their safety and effectiveness.

1.4 Optical imaging

Optical imaging offers a versatile approach to biomedical diagnostics, utilizing non-ionizing radiation to achieve imaging across various spatial and temporal scales (Solomon *et al* 2011). This technique covers a broad spectral range, from microwaves to ultraviolet (UV) light. However, due to the way biological tissues absorb and scatter light, deeper tissue penetration is most effective in the near-infrared (NIR) spectrum (700–1000 nm) (Wang 2007). By analyzing key tissue properties (absorption, scattering, and fluorescence) with the underlying physics of laser-tissue interactions, optical imaging enables the extraction of valuable structural, functional, and molecular data from targeted tissue regions. Further, when paired with contrast agents that are bio-conjugated, this technology can selectively enhance molecular targets, allowing for detailed visualization of biological processes in real-time. A key advantage of optical imaging is its ability to detect physiological changes at the molecular and cellular levels before anatomical abnormalities become apparent. This way these optical imaging technologies enable early detection capability and hence makes them powerful tools in disease diagnosis and biomedical research, with significant insights into tissue health and function.

1.5 Diagnostic multimodality imaging

Imaging of deep-tissue structures using optical techniques is limited by the optical diffusion limit, which typically extends only about 1 mm into the tissue (Wang 2007). Due to this, many conventional optical imaging techniques face a tradeoff between penetration depth and spatial resolution. Techniques such as those based on photoacoustic or ultrasound can achieve imaging depths of up to a few cm, but at the cost of spatial resolution (Weissleder and Pittet 2008), whereas optical coherence tomography (OCT)-like modalities and other time-gated techniques offer good spatial resolution (less than 10 μm), but imaging is possible only in thin or low-scattering tissue layers due to their shallow penetration depth (approximately 1 mm). In the case of high-resolution imaging methods, including confocal microscopy, provide detailed structural, functional, and molecular insights but have limited penetration depth (less than 0.2 mm).

From this perspective of tradeoff between penetration depth and resolution along with other attributes such a speed of imaging, there is a growing interest among the scientific community in the recent past for developing novel hybrid imaging technologies. They try to combine optical imaging with other energy forms, which can give the best of both worlds. A good example in this context is photoacoustic imaging, which integrates the best attributes of both ultrasound and optical imaging techniques. Photoacoustic imaging can attain good lateral resolution (<400 μm) and penetration depths of up to 6 cm (Wang 2007), enabling structural, functional, and molecular imaging. Table 1.2 shows the capabilities of different imaging modalities with respect to their spatial resolution and imaging depth (Rudin *et al* 1999, Phelps 2000, Beckmann *et al* 2001, Weissleder and Ntziachristos 2003, Weissleder and Pittet 2008, Ntziachristos *et al* 2005).

It is evident from table 1.2 that techniques such as MRI and nuclear-based imaging require extended imaging times, which generally limits their utility for real-time diagnostics. X-ray imaging provides high-resolution and faster imaging capabilities, but involves ionizing radiation, which restricts the frequency and amount of dose that can be applied. Although optical imaging techniques are capable of visualizing structural, functional, and molecular tissue states with different lateral and temporal resolutions, they often struggle with the depth of penetration and spatial resolution required for intended applications. Integrating

Table 1.2. Diagnostic imaging systems and their comparative analysis (Ntziachristos *et al* 2005, Joseph *et al* 2010, Joseph 2013, Perinchery *et al* 2019, Haridas *et al* 2020, Antony *et al* 2023).

Imaging modalities	Resolution	Depth	Additional notes
MRI	10–100 μm	No limit	Provides functional/molecular information
HSI	100 μm	Not specified	Provides structural, functional information
SIM/ES-SIM	100–300 nm	Not specified	Super-resolution microscopy. Provides structural information
X-ray CT	~50 μm	No limit	High anatomical detail, provides structural, functional, molecular information
Ultrasonic microscopy/ ultrasonic imaging	50–500 μm	mm–cm	Includes ultrasound-based methods, provides structural, molecular information
PAM/PAT	1–500 μm	mm–cm	Hybrid imaging (optical + acoustic), provides structural, functional, molecular information
Optical/ fluorescence nanoscopy	250 nm–2 μm	<1 mm	Super-resolution optical tools, provides structural, functional, molecular information

multiple imaging schemes in such context, with optical techniques as a primary component, has become a necessity to achieve adequate penetration depth and spatial resolution.

There are multimodal imaging systems that employ the systematic use of various imaging platforms to capture comprehensive multi-scale and multi-level information from a subject. This sequential multimodal imaging process requires the moving of the subject across different imaging systems. Examples of such systems include MRI-optical biopsy, XCT-optical biopsy, and XCT-ultrasound and then gather diverse types of data. This imaging approach works based on visual synthesis of the data from multiple modalities followed by powerful image registration techniques. Then, they co-register obtained images from different imaging equipment, significantly enhancing diagnostic capabilities (Slomka *et al* 2001, Ntziachristos *et al* 2002, Shekhar *et al* 2005).

Sequential multimodal imaging however, often encounters challenges, for accurately registering images from moving structures that undergo dynamic changes (Cherry and Dahlbom 2004). Hence, there was growing interest in integrating multiple imaging modalities into a single, cohesive unit. These synchronous multimodality imaging systems, which combine two or more imaging techniques into a unified setup, are designed to address some of the limitations of sequential approaches. They enable simultaneous capturing of object images thereby eliminating the issues caused by non-synchronous registration between different imaging methods.

Non-ionizing multimodal imaging systems started coming up to subdue the limitations of detrimental effects of using ionizing radiation, and also to improve the overall imaging performance in terms of resolution, contrast and depth of tissue imaging (Ntziachristos *et al* 2000). A contrast-enhanced diffuse optical tomography (DOT) of the breast and MR-guided optical spectroscopy (Ntziachristos *et al* 2002) were some of the initial reported systems in this regard. Both DOT and MRI have, however, faced challenges in terms of long imaging times (minutes to hours) and comparably low spatial resolution.

Obtained data from individual imaging modalities are to be integrated to provide complementary insights to obtain comprehensive multi-level and multi-scale information. Optical and ultrasound techniques, with their non-ionizing nature and scalable spatial and depth resolutions, can give real-time imaging capabilities. Among optical techniques, fluorescence microscopy and photoacoustic imaging are promising for surface and deep-tissue imaging, respectively (James *et al* 2014). Fluorescence microscopy can achieve sub-micron spatial resolution at the surface, while photoacoustic imaging can map optical absorption heterogeneities at mesoscopic resolution (James *et al* 2014, Lim and Murukeshan 2017). A comprehensive description and related aspects of this fascinating imaging modality is given in chapters 4 and 6. Ultrasound imaging, integrated with these techniques, can also provide complementary structural information at mesoscopic depth (James *et al* 2014). This multimodal approach offers unique capabilities for obtaining structural and molecular information across various spatial resolution scales.

Further, Raman spectroscopy and its latest embodiments are also started making significant impact in medical imaging. A detailed analysis of this technology is given from fundamentals to applications in chapter 5.

The following chapter will detail the basic optics concepts and optomechanical components and related equipment necessary for configuring such imaging modalities. This will be followed by a comprehensive coverage on diagnostic biomedical imaging, innovations in hyperspectral and photoacoustic imaging systems, and Raman spectroscopy, which are detailed based on the fundamental and applied principles.

1.6 Problems

1. What is multimodality imaging? What is its significance in medical diagnostics?
2. What is a synchronous multimodal imaging system? Give an example.
3. How do ionization radiation-based multimodal imaging systems work? Give three examples with details.
4. Name three high-resolution imaging techniques and do a comparative analysis of their performance in terms of resolution, cost and diagnostic sensitivity.
5. What is the principle of photoacoustic imaging? Is this an ionization radiation-based imaging?
6. How is photoacoustic imaging superior compared to confocal imaging or other optical imaging techniques?
7. Give a detailed write up on 'Why are optical imaging or hybrid optical imaging preferred for diagnostics applications?'
8. Refer to the latest literature on diagnostic medical imaging using hybrid optical techniques and perform a thorough comparison with current clinical imaging modalities.

References

Ale A, Ermolayev V, Herzog E, Cohrs C, de Angelis M H and Ntziachristos V 2012 FMT-XCT: *in vivo* animal studies with hybrid fluorescence molecular tomography-X-ray computed tomography *Nat. Methods* **9** 615–20

Antony M M, Haridas A, Suchand Sandeep C S and Vadakke Matham M 2023 An optodigital system for visualizing the leaf epidermal surface using embedded speckle SIM: a 3D non-destructive approach *Comput. Electron. Agric.* **211** 107962

Antony M M, Suchand Sandeep C S and Vadakke Matham M 2024 Hyperspectral vision beyond 3D: a review *Opt. Lasers Eng.* **178** 108238

Beckmann N, Mueggler T, Allegrini P R, Laurent D and Rudin M 2001 From anatomy to the target: contributions of magnetic resonance imaging to preclinical pharmaceutical research *Anat. Rec.* **265** 85–100

Beyer T T D, Brun T, Kinahan P E, Charron M, Roddy R, Jerin J, Young J, Byars L and Nutt R 2000 A combined PET/CT scanner for clinical oncology *J. Nucl. Med.* **41** 1369–79

Cherry S R and Dahlbom M 2004 *PET: Physics, Instrumentation, and Scanners* pp 1–124 (New York: Springer)

Cherry S R, Louie A Y and Jacobs R E 2008 The integration of positron emission tomography with magnetic resonance imaging *Proc. IEEE* **96** 416–38

Dinish U S, Gulati P, Murukeshan V M and Seah L K 2007 Diagnosis of colon cancer using frequency domain fluorescence imaging technique *Opt. Commun.* **271** 291–301

Haridas A, Perinchery S M, Shinde A, Buchnev O and Murukeshan V M 2020 Long working distance high resolution reflective sample imaging via structured embedded speckle illumination *Opt. Lasers Eng.* **134** 106296

Hill D L, Batchelor P G, Holden M and Hawkes D J 2001 Medical image registration *Phys. Med. Biol.* **46** R1–45

Hofmann M, Steinke F, Scheel V, Charpiat G, Farquhar J, Aschoff P, Brady M, Scholkopf B and Pichler B J 2008 MRI-based attenuation correction for PET/MRI: a novel approach combining pattern recognition and atlas registration *J. Nucl. Med.* **49** 1875–83

Joseph J, Sathiyamoorthy K, Visalatchi T *et al* 2010 Laser-induced photoacoustic spectroscopy investigation of colon phantom tissue *Appl. Phys.* A **101** 567–71

Joseph J 2013 Integrated multimodality imaging for tissue diagnostics *Doctoral thesis* (Singapore: Nanyang Technological University)

James J, Murukeshan V M and Woh L S 2014 Integrated photoacoustic, ultrasound and fluorescence platform for diagnostic medical imaging-proof of concept study with a tissue mimicking phantom *Biomed. Opt. Express* **5** 2135–44

Judenhofer M S *et al* 2008 Simultaneous PET-MRI: a new approach for functional and morphological imaging *Nat. Med.* **14** 459–65

Kalki K, Brown J K, Blankespoor S C, Hasegawa B H, Dae M W, Chin M and Stillson C A 1995 Myocardial perfusion imaging with a correlated X-ray CT and SPECT system: an animal study *IEEE Nuclear Science Symp. and Medical Imaging Conf. Record* **43** 2000–7

Lagarto J L, Villa F, Tisa S *et al* 2020 Real-time multispectral fluorescence lifetime imaging using single photon avalanche diode arrays *Sci. Rep.* **10** 8116

Li C, Yang Y, Mitchell G S and Cherry S R 2011 Simultaneous PET and multispectral 3-dimensional fluorescence optical tomography imaging system *J. Nucl. Med.* **52** 1268–75

Lim H-T and Murukeshan V M 2017 Hyperspectral photoacoustic spectroscopy of highly-absorbing samples for diagnostic ocular imaging applications *Int. J. Optomechatronics* **11** 36–46

Lim H T and Murukeshan V M 2016 A four-dimensional snapshot hyperspectral video-endoscope for bio-imaging applications *Sci. Rep.* **6** 24044

Lindsley E H, Nicolau D V, Wachman E S, Enderlein J, Leif R C, Farkas D L and Farkas D L 2004 The hyperspectral imaging endoscope: a new tool for *in vivo* cancer detection *Proc. SPIE* **11964** 1–7

Lu G and Fei B 2014 Medical hyperspectral imaging: a review *J. Biomed. Opt.* **19** 10901

Lucas A J, Hawkes R C, Ansorge R E, Williams G B, Nutt R E, Clark J C, Fryer T D and Carpenter T A 2006 Development of a combined microPET-MR system *Technol. Cancer Res. Treat.* **5** 337–41

MacDonald L R *et al* 2002 Evaluation of X-ray detectors for dual-modality CT-SPECT animal imaging *Proc. SPIE* **4786** 1

Malko J A, V H R, Gullberg G T and Kowalsky W P 1986 SPECT liver imaging using an iterative attenuation correction algorithm and an external flood source *J. Nucl. Med.* **27** 701–5

Matsuo K, Ota H, Akamatsu T, Sugiyama A and Katsuyama T 1997 Histochemistry of the surface mucous gel layer of the human colon *Gut* **40** 782–9

Marra K, LaRochelle E P, Chapman M S, Hoopes P J, Lukovits K, Maytin E V, Hasan T and Pogue B W 2018 Comparison of blue and white lamp light with sunlight for daylight-mediated, 5-ALA photodynamic therapy, *in vivo Photochem. Photobiol.* **94** 1049–57

Murukeshan V M and Sujatha N 2007 All fiber based multispeckle modality endoscopic system for imaging medical cavities *Rev. Sci. Instrum.* **78** 053106

Mohankumar V Krishnan, Matham M V, Krishnan S, Parasuraman P, Joseph J and Bhakoo K 2012 Red, green, and blue gray-value shift-based approach to whole-field imaging for tissue diagnostics *J. Biomed. Opt.* **17** 0760101

Nehmeh S A *et al* 2004 Quantitation of respiratory motion during 4D-PET/CT acquisition *Med. Phys.* **31** 1333–8

Neil M A, Juskaitis R and Wilson T 1997 Method of obtaining optical sectioning by using structured light in a conventional microscope *Opt. Lett.* **22** 1905–7

Ng T S, Bading J R, Park R, Sohi H, Procissi D, Colcher D, Conti P S, Cherry S R, Raubitschek A A and Jacobs R E 2012 Quantitative, simultaneous PET/MRI for intratumoral imaging with an MRI-compatible PET scanner *J. Nucl. Med.* **53** 1102–9

Ntziachristos V, Ripoll J, Wang L V and Weissleder R 2005 Looking and listening to light: the evolution of whole-body photonic imaging *Nat. Biotechnol.* **23** 313–20

Ntziachristos V, Yodh A G, Schnall M and Chance B 2000 Concurrent MRI and diffuse optical tomography of breast after indocyanine green enhancement *Proc. Natl Acad. Sci. USA* **97** 2767–72

Ntziachristos V, Yodh A G, Schnall M D and Chance B 2002 MRI-guided diffuse optical spectroscopy of malignant and benign breast lesions *Neoplasia* **4** 347–54

Perinchery S M, Haridas A, Shinde A, Buchnev O and Murukeshan V M 2019 Breaking diffraction limit of far-field imaging via structured illumination Bessel beam microscope (SIBM) *Opt. Express* **27** 6068–82

Phelps M E 2000 Positron emission tomography provides molecular imaging of biological processes *Proc. Natl Acad. Sci. USA* **97** 9226–33

Rudin M, Beckmann N, Porszasz R, Reese T, Bochelen D and Sauter A 1999 *In vivo* magnetic resonance methods in pharmaceutical research: current status and perspectives *NMR Biomed.* **12** 69–97

Schlemmer H P *et al* 2008 Simultaneous MR/PET imaging of the human brain: feasibility study *Radiology* **248** 1028–35

Shekhar R, W. V, Raja S, Zagrodsky V, Kanvinde M, Wu G and Bybel B 2005 Automated 3-dimensional elastic registration of whole-body PET and CT from separate or combined scanners *J. Nucl. Med.* **46** 1488–96

Slomka P J, Mandel J, Downey D and Fenster A 2001 Evaluation of voxel-based registration of 3-D power Doppler ultrasound and 3-D magnetic resonance angiographic images of carotid arteries *Ultrasound Med. Biol.* **27** 945–55

Solomon M, Liu Y, Berezin M Y and Achilefu S 2011 Optical imaging in cancer research: basic principles, tumor detection, and therapeutic monitoring *Med. Princ. Pract.* **20** 397–415

Szu H H, Gat N, Vetterli M, Campbell W J and Buss J R 2000 Imaging spectroscopy using tunable filters: a review *Proc. SPIE* **4056** 1–15

Tang H R, Brown J K and Hasegawa B H 1997 Use of X-ray CT-defined regions of interest for the determination of SPECT recovery coefficients *IEEE Trans. Nucl. Sci.* **44** 1594–9

Thomasson D M, Gharib A and Li K C 2004 A primer on molecular biology for imagers: VIII. Equipment for imaging molecular processes *Acad. Radiol.* **11** 1159–70

Townsend D W, Beyer T and Blodgett T M 2003 PET/CT scanners: a hardware approach to image fusion *Semin. Nucl. Med.* **33** 193–204

Townsend D W and Cherry S R 2001 Combining anatomy and function: the path to true image fusion *Eur. Radiol.* **11** 1968–74

Wang L V a H I W 2007 *Biomedical Optics: Principles and Imaging* (New York: Wiley)

Weisenberger A G *et al* 2003 SPECT-CT system for small animal imaging *IEEE Trans. Nucl. Sci.* **50** 74–9

Weissleder R and Ntziachristos V 2003 Shedding light onto live molecular targets *Nat. Med.* **9** 123–8

Weissleder R and Pittet M J 2008 Imaging in the era of molecular oncology *Nature* **452** 580–9

Wells P N 2006 Ultrasound imaging *Phys. Med. Biol.* **51** R83–98

IOP Publishing

Diagnostic Biomedical Optics

Fundamentals and applications

Murukeshan Vadakke Matham, C S Suchand Sandeep, Maria Merin Antony, Manojit Pramanik and Santhosh Chidangil

Chapter 2

Basic optics for diagnostic imaging

Murukeshan Vadakke Matham

Optics and related studies using optics use the properties and behaviour of light that include light–matter interaction. It also covers the light-based instruments that have variety of applications including imaging and diagnostics. This chapter hence gives a detailed description on some of the basic optics and optical components, light–matter interaction, basics of imaging, and related parameters, that are used for imaging applications. Different illumination beam profiles used for imaging purposes, such as Gaussian beams and Bessel beams will also be explained. Finally, some of the important aspects pertaining to the laser safety aspects will be discussed here.

2.1 Basic optics

The famous de Broglie hypothesis states that light shows the dual nature of particles as well as waves. This means it behaves as particles in some interactions but behaves as a wave in certain other situations (figure 2.1).

To understand this, let us consider the principles of geometrical optics. In this context, light is treated as a ray that travels in a straight line within each medium. This approach is referred to as ray optics, or alternatively, geometrical optics. The wave nature of light, based on Huygens' wave theory, was first proposed in the late 1600s. Around the same time, Newton introduced his particle theory of light. Both theories were able to explain the phenomena of reflection and refraction. A photon is the quanta of energy and light consists of many photons. The energy of the photon is given as

$$E = h\nu \tag{2.1}$$

E represents the energy of a photon, while h denotes Planck's constant (6.63×10^{-34} J s^{-1}). The symbol ν corresponds to the frequency of the light source, determining the photon's energy level.

doi:10.1088/978-0-7503-2364-2ch2

Figure 2.1. Particle and wave nature of light.

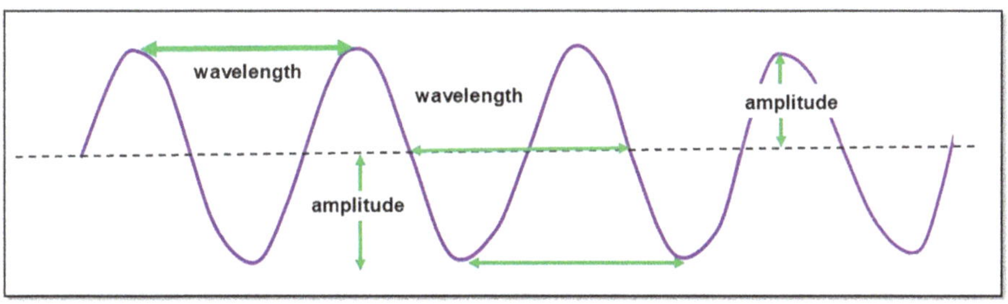

Figure 2.2. Representation of propagating wave.

$$\nu = c/\lambda \tag{2.2}$$

A standard depiction of a propagating wave, representing a photon, is illustrated in figure 2.2. This figure visually defines key wave properties, including wavelength and amplitude, providing a clear representation of their characteristics.

Law of reflection

The law of reflection states that the angle of reflection is equal to the angle of incidence. These angles are typically measured between the incident or reflected ray and a line perpendicular to the surface, known as the 'normal', at the point where the ray contacts the surface (see figure 2.3).

Snell's law

The Snell–Descartes law, commonly referred to as Snell's law, gives the relationship between the incidence angle and refraction angle when light or other waves pass through the boundary between two isotropic media (e.g., glass and air) (Born and Wolf 2020). This law plays a crucial role in ray tracing, enabling the determination

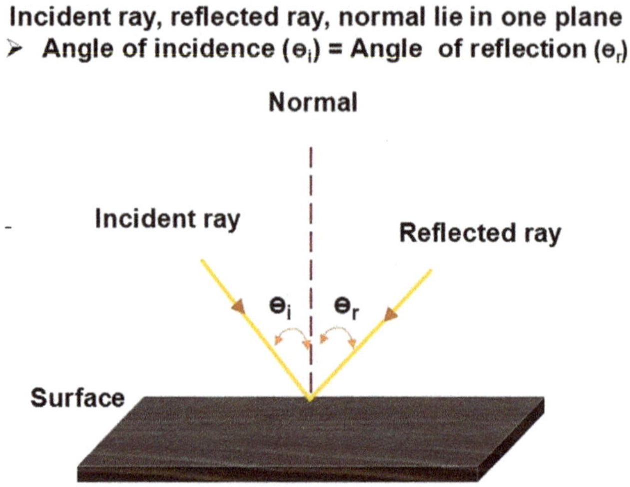

Figure 2.3. Pictorial representation of law of reflection.

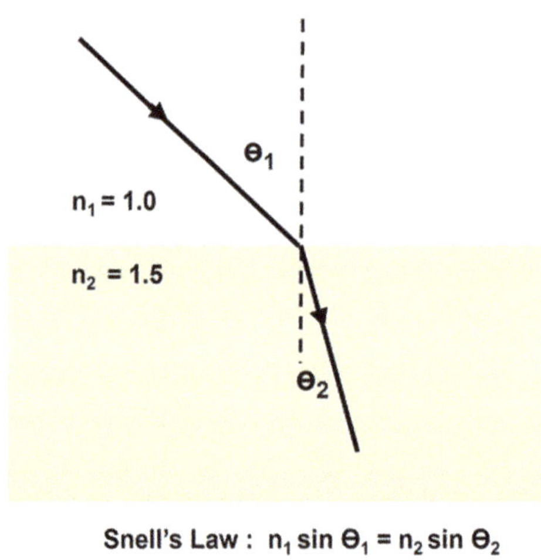

Figure 2.4. Pictorial representation of Snell's law of refraction.

of either the incidence angle or the refraction angle. It can also be used to calculate the refractive index of materials. Snell's law states that for a specific pair of media, the ratio between the sines of angles of incidence θ and refraction θ_2 is equal to the ratio of their respective medium refractive indices (n_2/n_1). It is also equal to the ratio of their phase velocities (v_1/v_2) as shown in equation (2.3).

$$\frac{\sin\theta_1}{\sin\theta_2} = \frac{n_2}{n_1} = \frac{v_1}{v_2} \tag{2.3}$$

where v_1, v_2 are velocities in the respective media (figure 2.4).

Total internal reflection and critical angle

When a light beam travels from a medium with a higher refractive index (n) to one with a lower refractive index (n), Snell's law suggests that in some instances (when the angle of incidence is sufficiently large), the sine of the angle of refraction exceeds one. The maximum angle of incidence that still results in refraction is known as the critical angle, where the refracted ray moves along the interface between the two media. If the angle of incidence surpasses this critical angle, the light does not transmit into the second medium but is completely reflected back into the original medium. This phenomenon is called total internal reflection (TIR). In figure 2.5, n and n' are represented as n_1 and n_2, respectively.

At the critical angle θ_c, the refracted ray travels parallel to the boundary. According to Snell's law,

$$n_1 \sin\theta_1 = n_2 \sin\theta_2 \tag{2.4}$$

$$n_1 \sin 90° = n_2 \sin\theta_c \tag{2.5}$$

$$\sin\theta_c = \frac{n_1}{n_2} \tag{2.6}$$

The critical angle, θ_c is defined as the angle of incidence at which the refracted ray emerges tangent to the surface. As shown in figure 2.5, when $\theta_a > \theta_c$, total internal reflection occurs, when light beam travels from denser medium (medium '**a**' with refractive index, n) to a rarer medium (medium '**b**' with refractive index, n').

Optical fibers, which are used for communication or imaging and sensing applications, uses total internal reflection (TIR) to confine and guide the light through the core of the fiber. An optical fiber consists of a central light-transmitting core, encased by a cladding that helps confine the light within the core as shown in figure 2.6. Generally, lasers such as Nd: YAG, femtosecond lasers, and many diode lasers that are used for imaging applications are pigtailed to optical fibers or coupled into optical fibers.

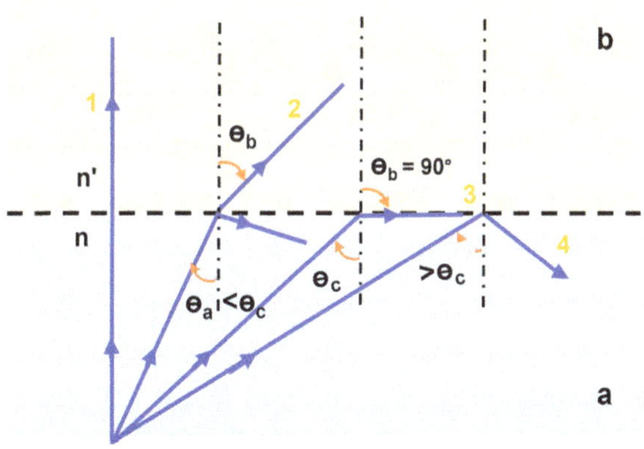

Figure 2.5. Illustration of total internal reflection.

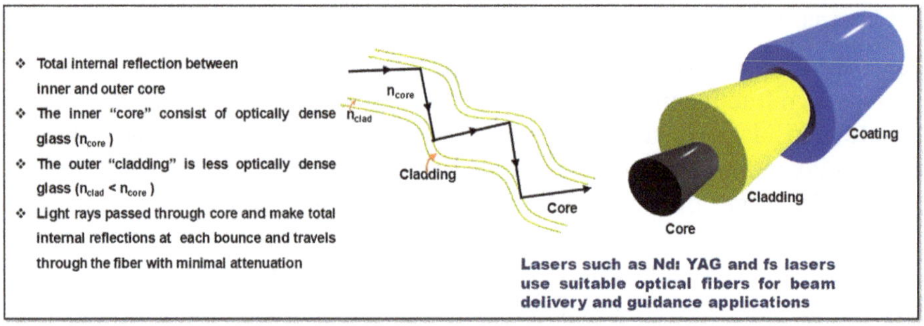

Figure 2.6. Light propagation in optical fibers.

2.2 Basic optical components

2.2.1 Laser

In general, a laser resonator consists of two parallel or (nearly) planar mirrors as shown in figure 2.7. One of these two mirrors will have 100% (close to 100%) reflectivity, the other mirror designed to have a reflectivity less than 100% (95% approx.) so that some of the light from the resonator can transmit through this mirror, which will act as the laser window for the photons to emit (Born and Wolf 2020).

Light travels back and forth between these two mirrors multiple times. They interfere constructively if the following equation is satisfied.

$$m\lambda_0 = 2L\cos\theta \tag{2.7}$$

$$m\lambda_0 = 2L \text{ for } \theta = 0^\circ \tag{2.8}$$

$$m\lambda_0/2 = L \tag{2.9}$$

Since light that travels perpendicular to the resonator mirrors will remain within the resonator, they will be significantly amplified. This will make a standing wave pattern between the forward to and backward travelling waves from these mirrors, and thus form standing wave patterns (modes), which are governed by the above equations.

If L is the cavity length, then L will be equal to an integral number of half-wavelengths ($\lambda/2$) for the laser to operate at a certain specific wavelength.

In terms of frequency, since $\nu = c/v_0$, the laser mode frequency is given by

$$\nu_m = mc/2L \tag{2.10}$$

These modes will have a separation between them with a frequency difference of $\Delta\nu = c/2L$, as shown in figure 2.8, where c is the velocity of light in vacuum. Hence, the spectrum of light of a laser looks like a series of narrow peaks spaced by the mode spacing.

Most laser systems use curved mirrors (see figure 2.9) instead of true planar mirrors as this can (i) make the laser alignment simple, and (ii) cavity losses get minimised.

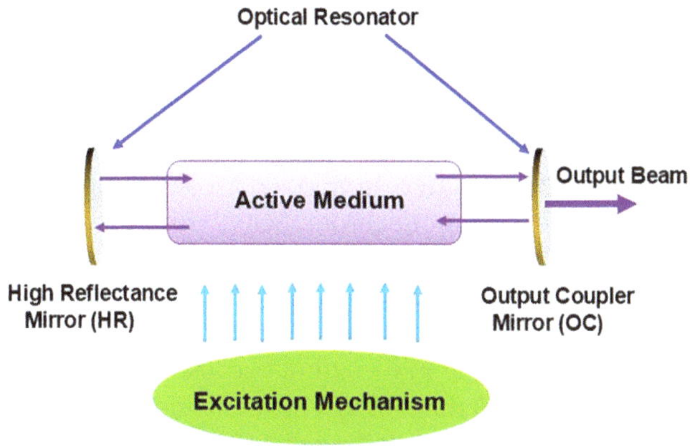

Figure 2.7. Basic components of a laser resonator.

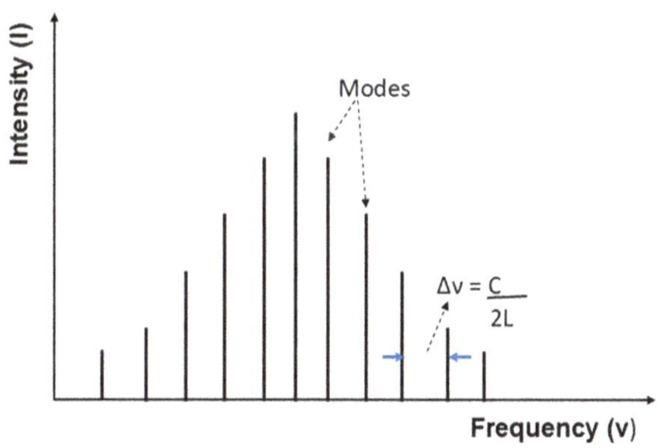

Figure 2.8. Longitudinal modes in a laser.

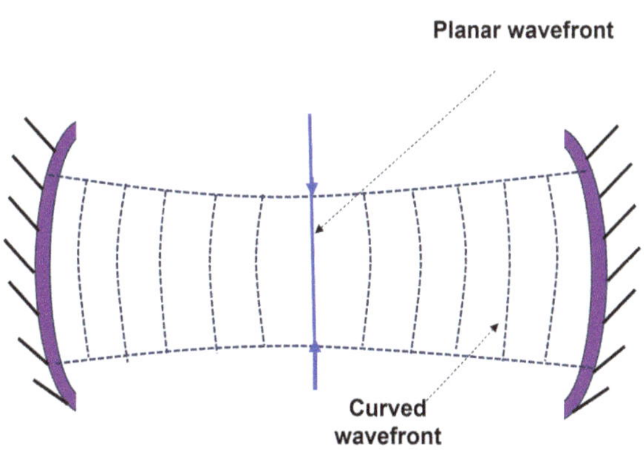

Figure 2.9. Curved resonators in a laser.

2.2.2 Laser beam: characteristics

LASER, the abbreviation for Light Amplification by Stimulated Emission of Radiation, is a coherent beam of electromagnetic radiation. The light propagates as a wave but also behaves as a particle of energy (photon) (Silfvast 2012). The laser, which is a coherent light source, differs from other broadband light sources due to its unique characteristics such as monochromaticity, directionality, and brightness apart from its high coherence.

2.2.2.1 Monochromaticity

The laser beam is monochromatic in its nature because of the narrow spectral lines compared to the conventional light source whose emission frequency covers a large bandwidth. A single mode emission can be achieved by making the laser oscillate in the fundamental TEM_{00}, which is known as Gaussian mode. The laser is light that has nearly a single colour (due to stimulated emission) in contrast with non-laser light sources (broadband light sources), which produce light mainly through spontaneous emission and gives out polychromatic light (see figure 2.10).

Monochromaticity is attributed to the fact that the gain occurs at a well-defined frequency, that is specific to the transition frequency of the atoms in the gain medium (Silfvast 2012). Furthermore, the laser light (a stimulated emission) is much narrower than the spontaneous emission from a single transition, as shown in figure 2.10.

2.2.2.2 Directionality

The laser emission is a stimulated emission, and emitted photons travels along the laser axis with perfect or near perfect collimation (all rays are parallel). The laser beam exhibits strong directionality, or collimation, due to its minimal divergence angle, typically ranging from 0.2 to 10 mrad. This property allows the beam to maintain its narrow profile over long distances, enabling precise focusing onto a small target area with minimal spread (Born and Wolf 2020).

2.2.2.3 Brightness

Spectral brightness plays a very important role when handling laser sources for many applications including imaging/diagnostic imaging and sensing. The spectral

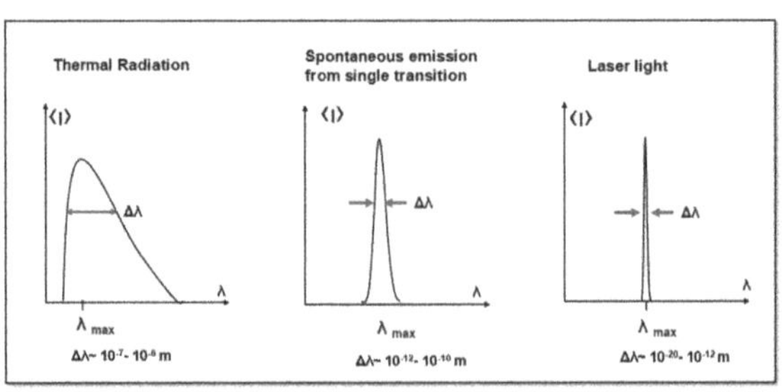

Figure 2.10. Spectral bandwidth comparison.

brightness of a He–Ne laser (1 mW) is ~10 000 000 000 times higher than that of the Sun (approx. 5 × 1023 W (m^3-sterad)$^{-1}$ (Born and Wolf 2020).

2.2.2.4 Coherence

The term, coherence is the measure of the correlation between the phases measured at different points on a propagating wave and is directly related to the wave source's characteristics.

The coherence of a laser beam is due to the fixed-phase relationship between two waves known as spatial coherence or between two points over a period of time of the same wave, which is known as temporal coherence. Temporal coherence how monochromatic the light source by measuring the correlation between the light wave phases at different points along the propagation direction, whereas spatial coherence is a measure of the correlation of the light wave phases at different points perpendicular to the propagation direction. This implies the uniformity of the wavefront's phase. Figure 2.11 gives an illustrative drawing of these two different types of coherence. (a) From coherent light source such as laser and (b) from an incoherent light source (Silfvast 2012, Born and Wolf 2020).

2.2.2.5 Divergence

A laser beam is highly directional, which means its divergence is very small. The directionality is described by the light beam divergence angle. Figure 2.12 illustrates the relationship between divergence and beam diameter in optical systems.

Perfect spatial coherent light with aperture diameter D will have unavoidable divergence because of diffraction (Silfvast 2012). From diffraction theory, the divergence angle θ_d is:

$$\theta_d = b\lambda/D \tag{2.11}$$

where λ and D are the wavelengths and the diameter of the beam, respectively. b is a coefficient whose value is approximately unity. It depends on the type of light

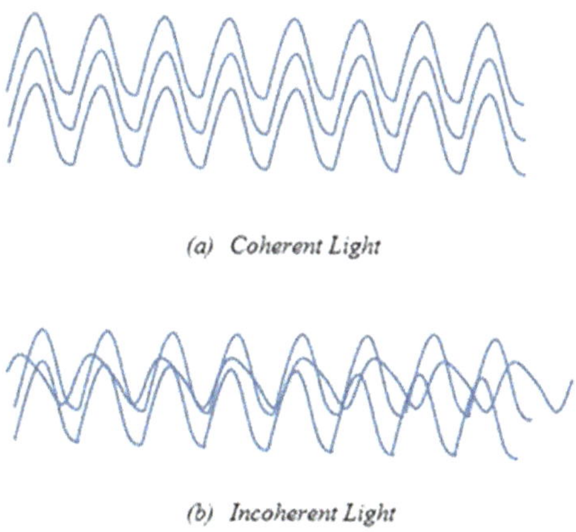

Figure 2.11. Coherent and incoherent light beams.

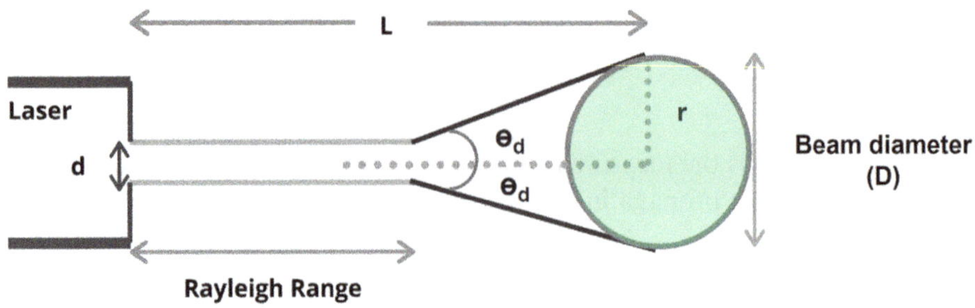

Figure 2.12. Laser beam divergence and directionality.

amplitude distribution and diameter of the beam. θ_d is called the diffraction limited divergence.

2.2.3 Laser beam: beam forming optics

2.2.3.1 Collimation optics

A collimator is a device designed to narrow a beam of particles or waves, allowing only those traveling parallel to a defined direction to pass through. Its function is to align the motion of the rays or particles in a specific direction. This makes the collimated light or parallel beam of rays that eventually causes the spatial cross section of the beam to become smaller and spread minimally when the beam propagates. There are many ways of implementing collimation. A diverging beam can be collimated if one keeps a lens at its focal plane from the origin of the source (see figure 2.13(a)).

In imaging, microscopy, and spectroscopy, maintaining a low beam divergence is essential, with an optimal requirement of approximately 2 mrad or less. A widely used and straightforward approach to achieving collimation involves using a single aspheric lens, as illustrated in figure 2.13(b). The focal length of this lens plays a critical role in determining the characteristics of the collimated beam—specifically, a longer focal length results in a larger beam diameter after collimation.

An alternative optical configuration utilizes a combination of lenses: a negative focal length lens paired with a positive lens (figure 2.13(c)). This setup offers greater flexibility, allowing for both collimation with beam expansion and collimation with beam reduction, depending on the specific optical requirements. Such configurations are particularly useful in applications where precise control over beam size and divergence is necessary.

If the laser beam is emanating from the end face of an optical fiber pigtailed to a laser, then the following configuration (figure 2.13(c)) can be employed for beam expansion and collimation.

2.2.3.2 Polarization

In an EM wave, by convention, the orientation of the electric field refers to the 'polarization' of light (Born and Wolf 2020). When light propagates through free space or an isotropic medium, it can be approximated as a plane wave, behaving as a

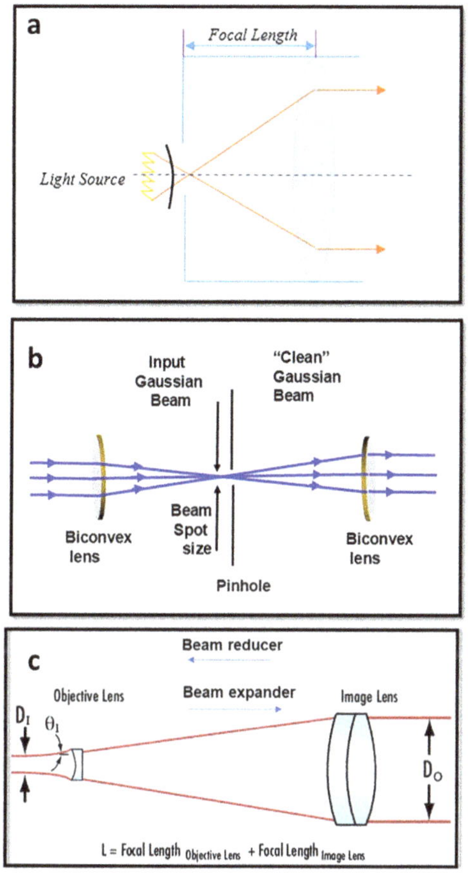

Figure 2.13. (a) Focussing of light using the lens, (b) collimation principle—with pinhole arrangement, (c) beam collimation, reduction, and expansion.

transverse wave in which the electric and magnetic fields oscillate perpendicular to the direction of propagation. The orientation of these oscillations defines the polarization state of the wave. In the case of linear polarization, the electric field oscillates in a fixed direction. However, when the field rotates at the optical frequency, the light exhibits circular or elliptical polarization. The direction of this rotation can be either clockwise or counterclockwise, a property known as the wave's chirality or handedness (see figure 2.14). This fundamental characteristic of polarization plays a crucial role in various optical applications, influencing interactions with materials and determining light's behaviour in complex optical systems.

2.2.3.3 Polarizers

The propagation of photons in space is characterized by its oscillatory motion in its electrical field vector, to produce an electromagnetic wave. This wave is representative of the direction and magnitude of the photon's electric field vector as a function

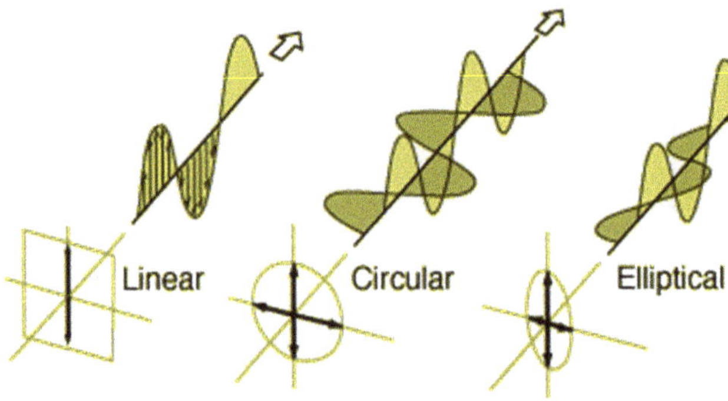

Figure 2.14. The different states of polarization of a beam.

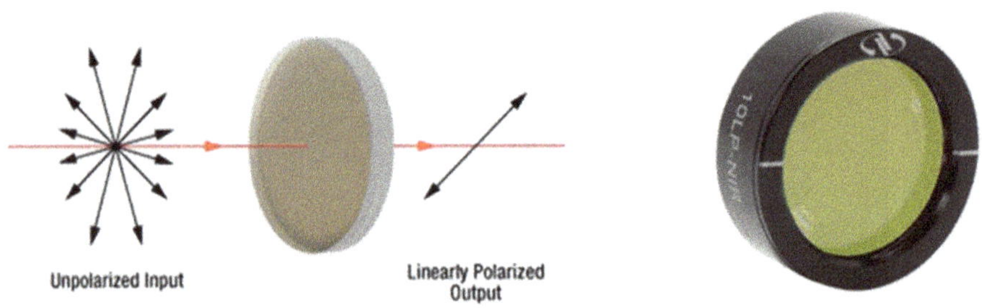

Figure 2.15. Linear polarizer.

of time. Due to the high coherence nature of a laser beam, all the photons are aligned in the same direction and thus the laser beam can be defined as linearly polarized.

Used as a common optical material, glass is isotropic and preserves the wave polarization. However, in certain materials known as birefringent materials, a wave's polarization will generally be modified when light propagates through them. A polarizer is an optical filter that transmits only one polarization (see figure 2.15). It converts a light beam of mixed/undefined polarization into a polarized light beam, which would have a well-defined polarization. The most common types of polarizers are linear and circular polarizers.

The two different categories of linear polarizers are absorptive and beam-splitting polarizers. In an absorptive polarizer, all the unwanted polarization states are absorbed, while for a beam-splitting polarizer, the unpolarized beam is divided into two with opposite polarization states.

Malus' law says that when a perfect polarizer is placed in a polarized beam of light, the intensity, I (where $I = E^2$), of the light that passes through is given by (Born and Wolf 2020),

$$I = I_0 \cos^2 \theta p \tag{2.12}$$

where I_0 represents the initial intensity of the light, and θp denotes the angle between the initial polarization direction of the light and the axis of the polarizer, as illustrated in figure 2.16.

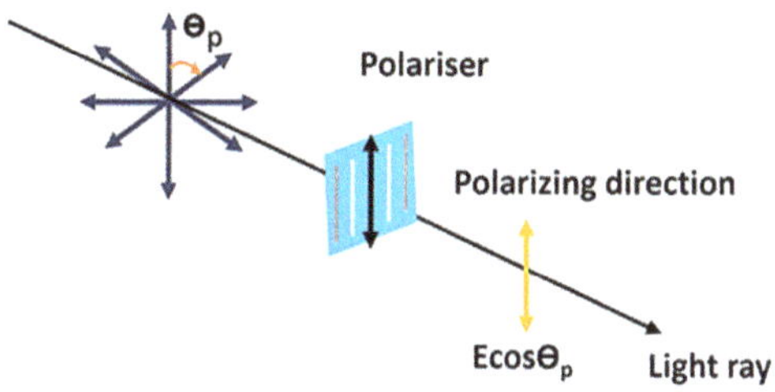

Figure 2.16. Malus law illustration.

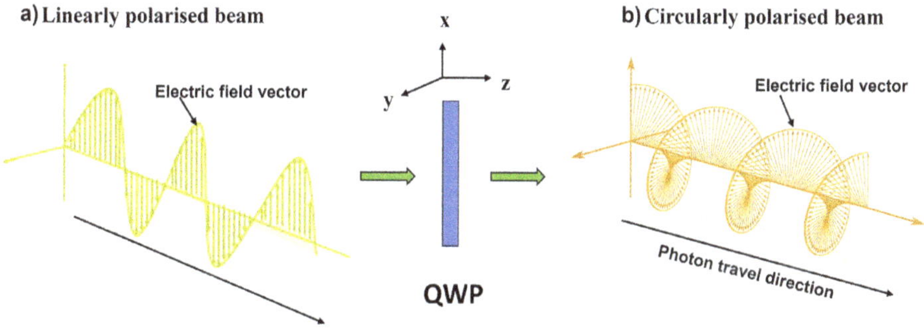

Figure 2.17. Illustration of working principle of QWP.

2.2.3.4 Wave plates

A wave plate, also known as a retarder, is an optical component used to modify the polarization state of a light beam as it passes through. The two most widely used types are the half-wave plate (HWP) and the quarter-wave plate (QWP) (Born and Wolf 2020). A half-wave plate shifts the polarization states of linearly polarized light, and a quarter-wave plate converts linearly polarized light into circularly polarized light as shown in figure 2.17.

Wave plates are made out of birefringent materials such as quartz or mica. Birefringent materials have a different index of refraction in different orientations with respect to the light passing through it. The behaviour of a wave plate, be it half wave or a quarter wave, depends on the crystal's thickness, the wavelength of the light used, and the variation of index of refraction. A suitable selection of the specified parameters allows for a controlled phase difference to be introduced between the two polarization components of a light wave, thereby leading to a change in the state of polarization.

The state of polarization of light has detrimental effects when light interacts with matter. For example, circular polarization has no directional effects whereas linear polarization does. An example to illustrate this is shown below in figure 2.18. It shows the effects of the state of polarization of the impinging beam when a femtosecond laser beam is used to machine holes on Au–Cr coated quartz plate. Such effects can be observed when laser interacts with bio surfaces as well.

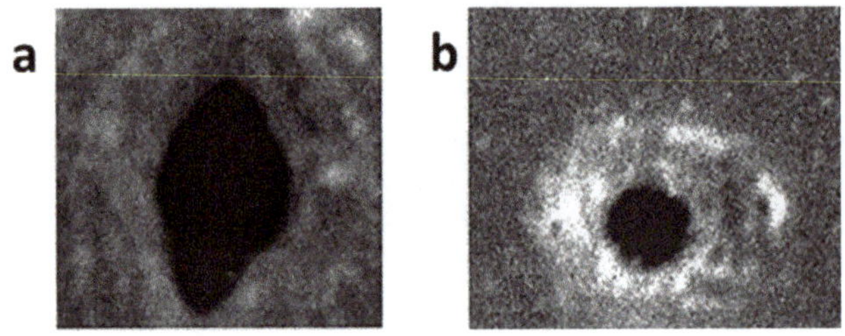

Figure 2.18. Effect of state of polarization. (a) Linearly polarized and (b) circularly polarized.

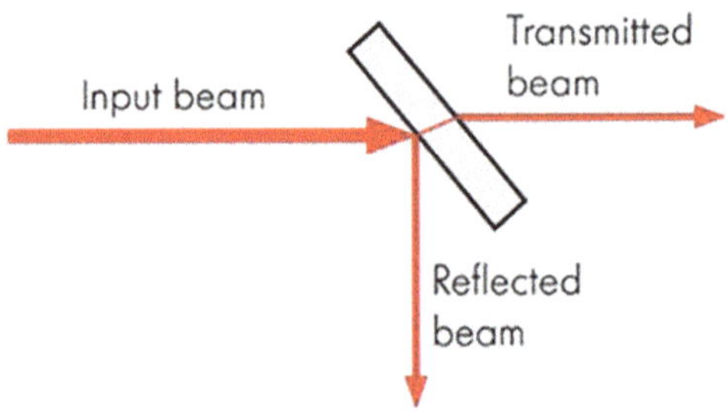

Figure 2.19. A 50:50 beam splitter configuration.

2.2.3.5 Beam splitters

A beam splitter is an optical component that splits light beam into two (see figure 2.19). They are widely used in most of the interferometers and form a crucial segment. In general, common types of beam splitters consist of a cube formed by joining two triangular glass prisms at their base using adhesives like polyester, epoxy, or urethane. The thickness of the adhesive layer is calibrated so that, for a specific wavelength of light, half of the light entering through one face of the cube is reflected, while the other half is transmitted. This happens due to frustrated total internal reflections of the light in the thin epoxy layer. Polarizing beam splitters, such as the Wollaston prism, split the light beam into two beams of different polarizations. This is achieved by the use of birefringent materials in its construction (figure 2.20).

Other possible designs of a beam splitter use a half-silvered mirror, a sheet of glass (figure 2.19(c)), or plastics with a transparent (thin) coating of a metal. The thickness of the coating is decided such that a portion of the incident light (at a 45° angle) is not absorbed by the coating and gets transmitted. The remainder of the light is thus reflected. This type of half-silvered mirrors is also known as 'pellicle mirror'.

A dichroic optical coating is used for certain special cases especially for selective transmission or reflection of different wavelengths as shown in figure 2.21. The ratio of reflection to transmission varies depending on the properties of the material and changes as a function of the incident light's wavelength.

Another type of beam splitter that is used in imaging instruments and applications, is a dichroic mirrored prism assembly. This optical component uses a dichroic

Figure 2.20. Different beam splitting optical components. (a) Glass plate beam splitter; (b) cube beam splitter, and (c) pellicle beam splitter.

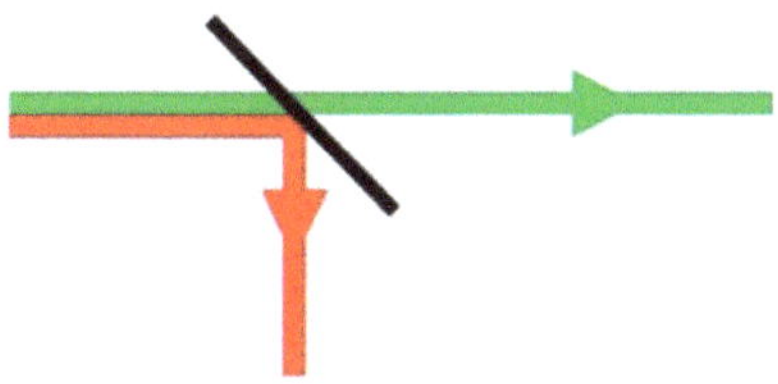

Figure 2.21. A dichroic beam splitter.

optical coating to split and divide an incoming light beam into a number of spectrally distinct output beams (see figure 2.22).

These types of dichroic beam splitters and mirrors are used in many imaging systems, spectrometers and imaging spectroscopes.

2.2.4 Spectrometer

A spectrophotometer consists of several fundamental components, including a light source, a sample holder, a wavelength dispersion element, and a detector, as illustrated in figure 2.23. Light passes into the spectrometer through an entrance slit. The slit helps to control the amount of light entering the instrument, ensuring the correct resolution and clarity of the spectrum.

The spectrometer works by dispersing light into its constituent wavelengths using a combination of mirrors and a diffraction grating. Light first enters through an entrance slit, which controls the beam size and ensures it is spatially constrained. This light is then directed toward a collimating mirror, which converts the divergent light into a parallel beam. The parallel light is reflected onto a diffraction grating, which separates the light into its various wavelengths based on diffraction principles, where different wavelengths are dispersed at different angles. A focusing mirror then collects the diffracted light and focuses it onto the exit slit, allowing only a specific wavelength (or a narrow range of wavelengths) to pass through. By rotating the diffraction grating, different wavelengths can be selected for analysis. This setup provides high-resolution spectral separation, commonly used in spectrometers and monochromators.

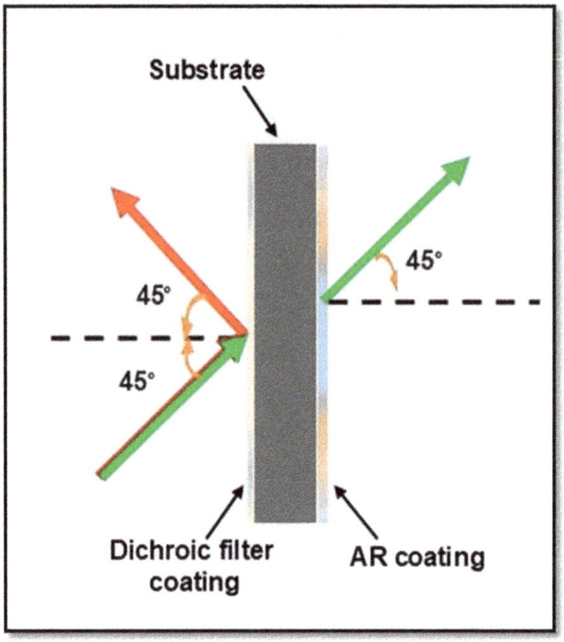

Figure 2.22. Working principle of a dichroic beam splitter.

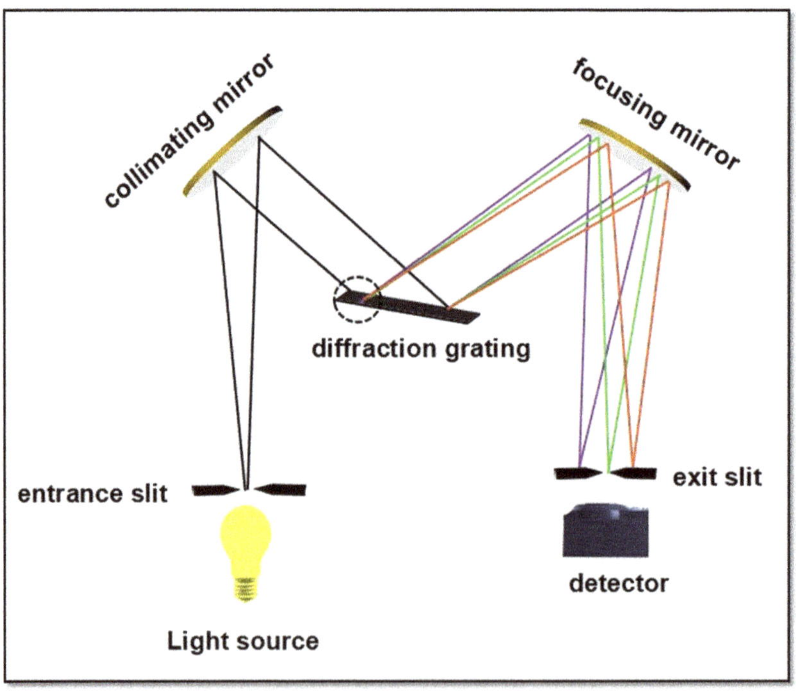

Figure 2.23. Schematic diagram of the spectrometer.

2.3 Illumination beam profiles

2.3.1 Gaussian beam

A laser beam is considered Gaussian if its irradiance profile follows a perfect Gaussian distribution (Silfvast 2012). In practice, most laser beams exhibit some degree of deviation from this ideal behaviour. The beam quality factor, or M^2 factor, measures how closely the beam's performance aligns with that of a diffraction-limited Gaussian beam. Gaussian irradiance profiles are symmetric about the beam's center and gradually decrease as the distance from the center, perpendicular to the beam's direction of propagation, increases (see figure 2.24). This behaviour is mathematically represented by equation (2.12) (Born and Wolf 2020).

$$I(r) = I_0 \exp\left(-\frac{2r^2}{w(z)^2}\right) = 2P/\pi w(z)^2 \exp\left(-\frac{2r^2}{w(z)^2}\right) \tag{2.12}$$

In equation (2.12), I_0 denotes the maximum irradiance at the beam's center, r is the radial distance from the beam's axis, $w(z)$ represents the beam radius at which the irradiance is $1/e^2$ of I_0, z is the distance travelled from the plane where the wavefront is flat, and P is the total power of the beam.

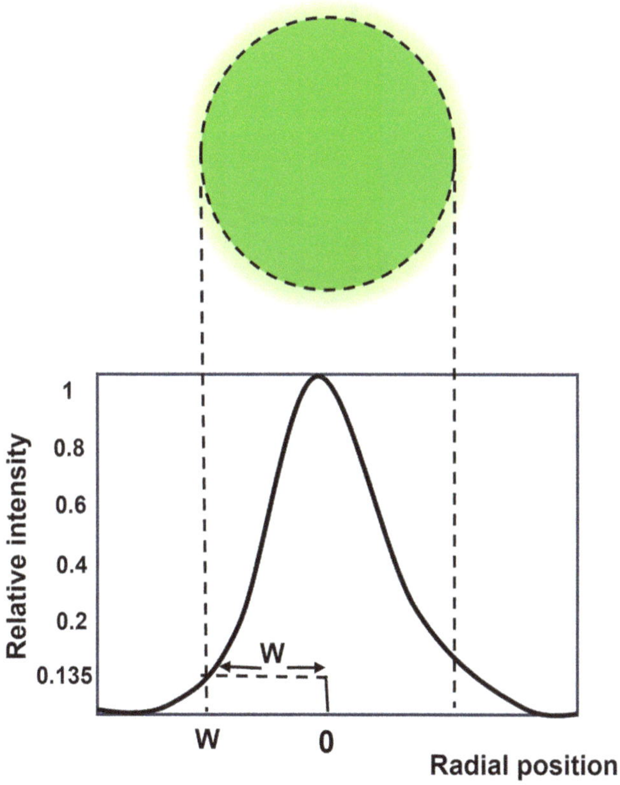

Figure 2.24. Characteristics of a Gaussian beam.

However, the irradiance profile changes as the beam propagates through space, which is why $w(z)$ depends on z. Due to diffraction, a Gaussian beam converges and then diverges from a region known as the beam waist (w_0), where the beam's diameter is at its smallest. The beam converges and diverges symmetrically on either side of the beam waist at an angle of divergence θ_d (figure 2.25). The relationship between the beam waist and the divergence angle is described by equations (2.13) and (2.14).

$$w_0 = \lambda/\pi\theta_d \tag{2.13}$$

$$\theta_d = \lambda/\pi w_0 \tag{2.14}$$

Here, λ represents the laser wavelength, and θ_d is the far-field approximation of the divergence angle, as depicted in the figure. According to equation (2.14), a smaller beam waist corresponds to a larger divergence angle, while a larger beam waist results in a smaller divergence angle, indicating a more collimated beam. This explains the use of beam expanders, which increase beam diameter to reduce divergence.

The beam diameter variation in the beam waist is defined by:

$$w(z) = w_0\sqrt{1 + \left(\frac{\lambda z}{\pi w_0^2}\right)^2} \tag{2.15}$$

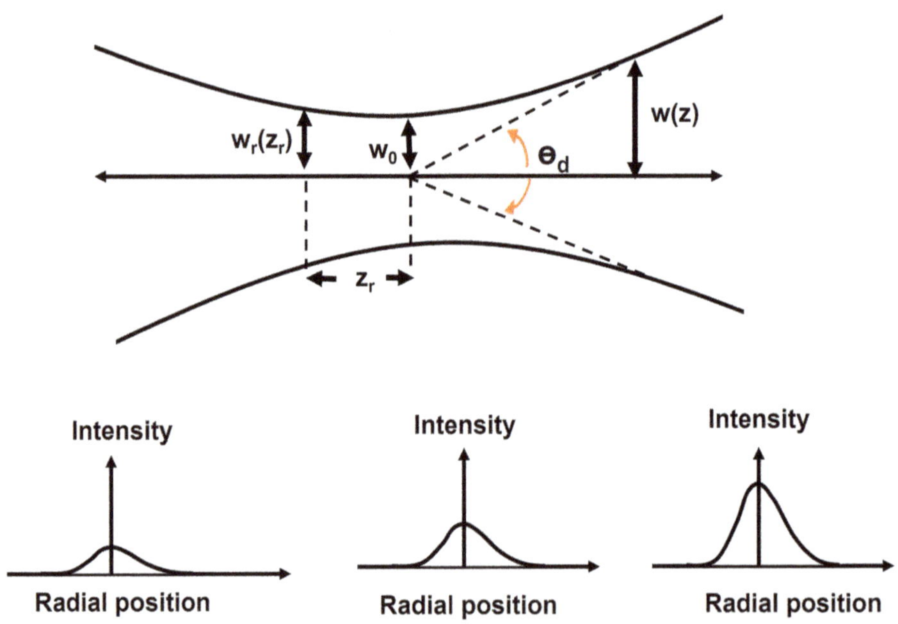

Figure 2.25. Gaussian beams characterized by key parameters, including beam waist (w_0), Rayleigh range (z_R), and divergence angle (θ_d).

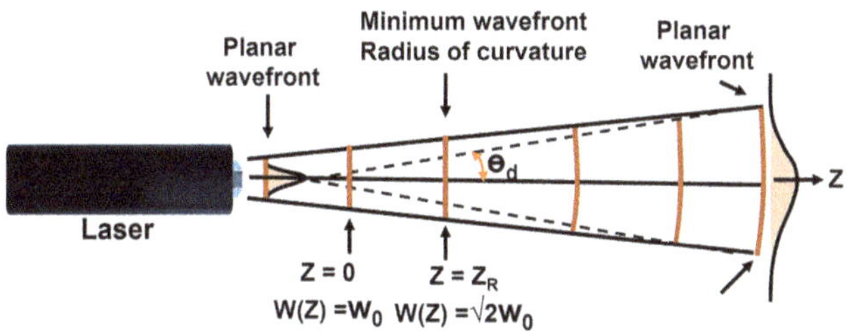

Figure 2.26. Gaussian beam propagation—illustration of beam waist and wave curvature.

The Rayleigh range of a Gaussian beam is the distance along the z-axis where the beam's cross-sectional area doubles in size. This happens when $w(z)$ increases to $\sqrt{2}$ times the beam waist, w_0. Based on equation (2.15), the Rayleigh range (z_R) can be represented as (Silfvast 2012):

$$z_R = \frac{\pi w_0^2}{\lambda} \tag{2.16}$$

This allows $w(z)$ to also be related to z_R:

$$w(z) = w_0\sqrt{1 + \left(\frac{\lambda z}{\pi w_0^2}\right)^2} = w_0\sqrt{1 + \left(\frac{z}{z_R}\right)^2} \tag{2.17}$$

Starting from the beam waist, the radius of curvature decreases from infinity to a minimum at the Rayleigh range. And as the beam moves further from the laser it returns to infinity (figure 2.26), occurring symmetrically on both sides of the waist.

At the beam waist, the laser's wavefront is flat and gradually returns to this shape as the distance from the waist increases. This happens because the wavefront's radius of curvature approaches infinity.

2.3.2 Bessel beam

Generally, Bessel beams are non-diffractive, meaning they do not spread out as they propagate. Additionally, Bessel beams are self-healing, allowing them to reform after partial obstruction further along the beam axis. A perfect Bessel beam is unbounded and require infinite energy and hence cannot normally exist. However, practical approximations can be made, which are valuable in many optical applications due to their minimal diffraction over a limited range. These approximations are usually obtained by directing a Gaussian beam through an axicon lens to produce a Bessel–Gauss beam, employing axisymmetric diffraction gratings, or positioning a small annular aperture in the far field. Additionally, higher-order Bessel beams can be created using spiral diffraction gratings.

2.3.2.1 Application of Bessel beam to microscopy

Bessel beam microscopy is an advanced plane illumination technique that captures images in a wide-field mode while incorporating structured illumination approaches, including two-photon (TP) excitation. It primarily operates in three commonly used modes: plane illumination, structured plane illumination, and TP plane illumination. The core principle of this imaging method involves the use of an axicon, a specialized optical component that modifies the point spread function (PSF) to generate a Bessel-like beam. In addition to the axicon, the system includes a lens positioned at its focal length from the image plane and a camera placed at a predetermined distance from the axicon.

This microscopy technique offers high spatial resolution and minimizes photodamage, though these parameters vary across different modes. While all three modes maintain similar spatial depth, plane illumination mode provides the highest imaging depth along the z-axis but has the lowest optical sectioning capability. In contrast, TP plane illumination mode delivers the best optical sectioning performance, the highest imaging speed, and the lowest photodamage, similar to the standard plane illumination mode. However, TP plane illumination is restricted to black-and-white imaging, whereas the other modes support colour imaging. Structured illumination mode balances several performance aspects but tends to induce higher photobleaching compared to the other two modes.

2.3.2.2 Advantages of Bessel beam

The main advantage of Bessel beam microscopy is its ability to capture dynamic cellular processes in real time. It can acquire images at several hundred frames per second (fps) and record up to 40 planes per second. The thin and fast-moving light sheets used in Bessel beam microscopy reduce photobleaching and phototoxicity, allowing for long-term imaging of live cells with minimal damage (Perinchery *et al* 2019, Hong *et al* 2020). Furthermore, this technique surpasses conventional high-resolution microscopy by providing near-isotropic spatial resolution while effectively minimizing photodamage and photobleaching effects.

A notable limitation of many traditional imaging methods, including standard light-sheet microscopy, is that only the side of the specimen aligned with the objective lens is illuminated. Additionally, thick samples often cause beam scattering along the z-axis, reducing image clarity. Bessel beam microscopy overcomes this issue by lateral scanning in the focal plane, allowing for improved illumination. A unique feature of Bessel beams is their self-healing property, where scattered photons reconstruct through constructive interference. This self-healing effect significantly enhances image quality and depth penetration, making Bessel beam microscopy a powerful tool for high-resolution, live-cell imaging.

2.3.3 Bessel–Gauss beams

Bessel–Gauss beam can be generated by passing through an axicon lens. Unlike a conventional spherical lens, which concentrates light at a single focal point, an axicon lens has an extended focus along the propagation axis, which increases the depth of focus. As the energy is spread over multiple points along the optical axis, its distribution leads to a reduction in both optical power and image contrast. Under on-axis illumination, the intensity distribution of the resulting Bessel–Gauss beam can be mathematically approximated as follows (Herman and Wiggins 1991):

$$I(r,\, z) = \frac{4\pi^2 E^2(R_{\text{ill}})}{\lambda}\frac{R_{\text{ill}} \sin \beta}{\cos^2 \beta} J_0^2\left(\frac{2\pi r \sin \beta}{\lambda}\right), \tag{2.18}$$

where $E^2(R_{\text{ill}})$ represents the incident beam's energy, R_{ill} refers to the illumination beam's radius of the axicon lens, and J_0 is the Bessel–Gauss beam. The axicon angle γ, β, and depth of focus Z_D, are related as shown in the equations given below:

$$n \sin \gamma = \sin (\gamma + \beta), \tag{2.19}$$

and

$$Z_D = R_{\text{ill}}(\cot \beta - \tan \gamma), \tag{2.20}$$

where n is the refractive index of the axicon lens. For small values of γ, Z_D can be simplified as:

$$Z_D = \frac{R_{\text{ill}}}{(n - 1)\gamma}. \tag{2.21}$$

A possible Gaussian beam geometry focussing to a point using a spherical lens is shown in figure 2.27(a). Figure 2.27(b) shows the optical configuration of a J_0 beam production. It can be noted that it uses a spherical lens with an annular pupil at the back focal plane for Bessel beam generation. In this system, the annular pupil also acts as a beam stop, restricting the amount of light that reaches the lens. As shown in figure 2.27(c), an axicon lens generates a J_0 beam, focusing light over an extended distance along the propagation axis, which enhances the depth of field.

2.4 Light sheet fluorescence imaging using Bessel–Gauss beam

We know that static light sheet configurations require lower peak intensities to reduce photodamage. In the case of virtual or digitally scanned light sheets, by configuring properly, we can provide greater control over the intensity profile and illumination sheet size ensuring uniform illumination across each line of the sample. This allows for the quantitative study of large samples in their natural and dynamic states. It is to be mentioned that despite using higher laser power, the digitally scanned light sheet does not cause increased photobleaching, as it operates within a linear photobleaching regime. Additionally, the total illumination energy required to produce images of equal brightness remains unchanged (Shinoj *et al* 2016, Hong *et al* 2017, Perinchery *et al* 2019). Figure 2.28 illustrates the distinction between static and virtual light sheets.

Though the detection axis optical components of digitally scanned light sheet microscopy resemble those in static light sheet microscopy, the illumination axis requires additional elements. These include a plano-convex axicon lens for generating the Bessel–Gauss beam, a collimation lens for focusing, a two-axis galvanometer for beam scanning, and a 4F system comprising a visible scan lens and an infinity-

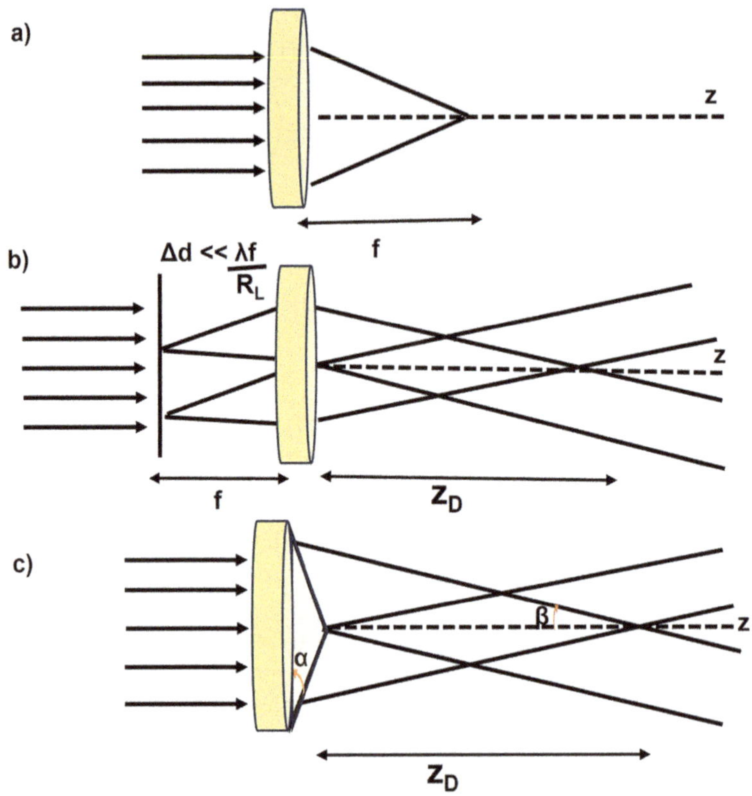

Figure 2.27. Illustration of focusing with (a) spherical lens, (b) lens and annular pupil, and (c) axicon lens.

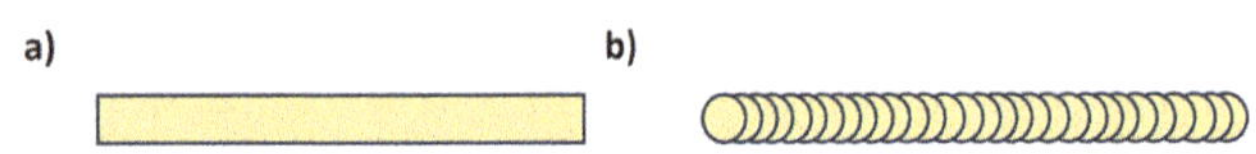

Figure 2.28. Diagrammatic representation of (a) static light sheet and (b) digitally-scanned light sheet.

corrected tube lens to relay the scan plane to the illumination objective's back aperture. Additionally, a long working distance objective lens replaces the cylindrical lens.

To overcome the limitations of Gaussian illumination, an axicon lens is used for on-axis illumination, efficiently generating a Bessel–Gauss beam. This approach is superior to using a finite-width annular pupil, as it eliminates the need for a beam stop, enhancing illumination efficiency, reducing exposure times, and minimizing photobleaching and photodamage. Beyond the beam's depth of focus (Z_D) otherwise known as beam length, the light rays expand into a ring. The 4F configuration's scan and tube lenses alternate the beam between its beam phase and ring phase, as shown in figure 2.29 (Perinchery *et al* 2016, Shinoj *et al* 2016, Suchand Sandeep *et al* 2022, Hong *et al* 2017, Hong 2018). The Z_D, in this setup is controlled by a variable

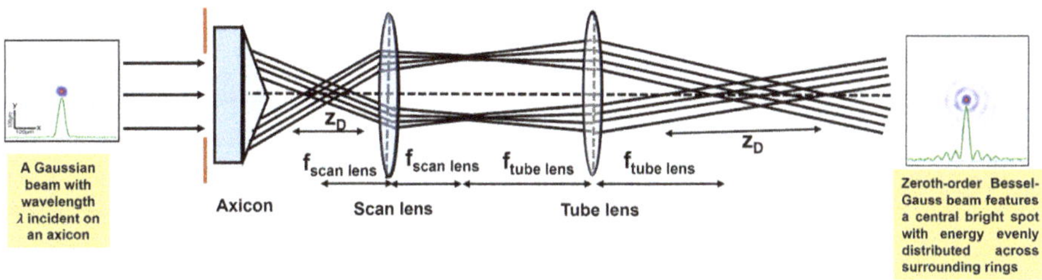

Figure 2.29. Bessel–Gauss beam—Generation and propagation along with Zemax simulations.

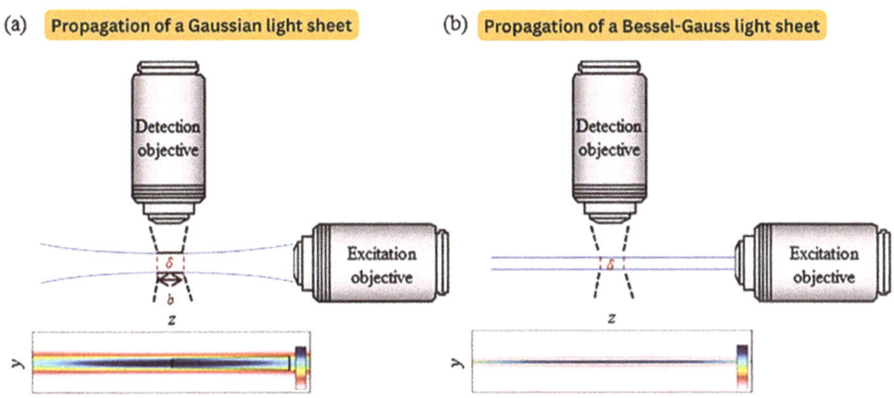

Figure 2.30. (a) Gaussian light sheet propagation and (b) Bessel–Gauss light sheet propagation.

aperture before the axicon, and its value can be estimated by equation (2.21). For a given numerical aperture (NA_{ill}), the larger the aperture radius (R_{ill}), the longer the beam length Z_D, with more energy concentrated in the side lobes. This means the Bessel-like properties and beam length are linear with the aperture radius.

The thickness of the virtual light sheet is directly proportional to $\frac{\lambda_0}{2NA_{ill}}$, where NA_{ill} is the numerical aperture of the illumination objective. A crucial aspect of this imaging technique is ensuring that the projected beam effectively and precisely covers the region of interest. Failure to do so may result in increased energy in the side lobes, leading to higher excitation tails on either side of the central core. This, in turn, degrades axial resolution and negatively impacts image quality.

The Galvano mirror with sweeping action directs the focused Bessel–Gauss beam along the y-axis, generating a digitally scanned or virtual light sheet for fluorescence excitation. This light sheet is within the detection objective lens's depth of field (DOF), as illustrated in figure 2.30. The figure also contrasts the relative intensity cross-section and radial intensity of a digitally scanned Gaussian beam with those of a Bessel–Gauss beam. Zemax® simulations indicate that the Bessel–Gauss beam offers a longer depth of focus and a narrower central peak than the Gaussian beam, effectively balancing the trade-off between light sheet thickness and length. To enhance anatomical distinction and image contrast, the fluorescence threshold must surpass the peak intensities of the side lobes. The 2D optical sections, featuring a broad field of view (FOV), are made possible by confining irradiation to the observed plane while simultaneously imaging fluorescence across the entire excitation plane. Fast image acquisition while minimizing photodamage is then attained by proper use of the sensitive camera that are going to be integral part of the optical set up. The imaging setup of a possible Bessel–Gauss light sheet fluorescence imaging is shown in figure 2.31(a.)

Building upon the discussion of Bessel–Gauss beam scanned light sheet fluorescence microscopy, this technique has been successfully employed by researchers/clinicians to image the trabecular meshwork of the eye with remarkable precision (Suchand Sandeep *et al* 2019). One of its key advantages is its optical sectioning capability, which enables the reconstruction of high-resolution, 3D volumetric images of the trabecular meshwork (TM) in an eye. This non-invasive approach eliminates the need for physical dissection, preserving the structural integrity of the tissue while providing valuable insights into its morphology and function.

The optical sections of the TM obtained through the experimental setup depicted in figure 2.31(a) serve as the foundation for its three-dimensional reconstruction (Suchand Sandeep *et al* 2019). To ensure accuracy, these optically sectioned images undergo a preprocessing step to correct any translational shifts that may result from objective lens movement. This correction is performed using the StackReg plugin in ImageJ, which employs an automatic pyramidal approach sub-pixel registration algorithm. The algorithm aligns each image slice progressively by designating a global anchor slice as a reference. Each subsequent slice is then aligned to this reference, and this process continues iteratively, ensuring a well-aligned image stack. Once the alignment is completed, image contrast adjustments are made, followed by background subtraction using the rolling-ball algorithm. In this method, the local background value for each pixel is determined by averaging over a large ball that rolls across the intensity surface surrounding the pixel. This background value is then subtracted from the image, effectively eliminating uneven variations in background intensities and enhancing contrast.

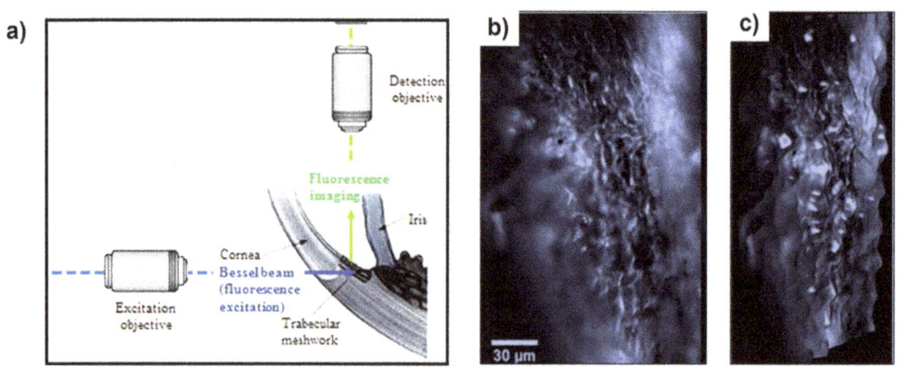

Figure 2.31. (a) Schematic diagram of Bessel–Gauss light sheet fluorescence imaging for eye imaging (b) Extended depth of focus image of the TM constructed from the optical sections. c) 3D visualization of TM.

For three-dimensional visualization of the TM, a combination of complex wavelet transform-based and image formation model-based extended depth of field (EDF) algorithms is utilized (Suchand Sandeep *et al* 2019). Figure 2.31(b) illustrates the 3D structure generated using the EDF algorithm, while figure 2.31(c) presents a structure-based 3D visualization reconstructed through the EDF algorithm plugin in ImageJ. This detailed 3D representation of the TM provides valuable insights for clinicians, aiding in the assessment of potential causes of open-angle glaucoma and informing the development of optimal treatment strategies (Suchand Sandeep *et al* 2019). Despite its promising applications, the *in vivo* implementation of this technique presents certain challenges. To facilitate *in vivo* imaging using the proposed system, a medical-grade fluorescein sodium solution can be applied as eye drops for fluorescent labelling of the TM. The fluorescein molecules diffuse through the cornea into the anterior chamber and exit via the iridocorneal angle (ICA), staining the TM in the process. Faster and more localized staining of the TM can be achieved through direct injection of fluorescein into the anterior chamber. However, these methods are not ideal for routine clinical investigations due to their invasive nature and the transient retention of fluorescein within the blood or anterior chamber. The use of fluorescein eye drops offers a more practical and non-invasive alternative, allowing for an extended imaging period of the TM. This technique is commonly employed in clinical settings for eye pressure evaluations and can be seamlessly integrated into the proposed imaging system. Given its affordability and ease of application, fluorescein eye drop administration represents a viable approach for translating this imaging method into clinical practice.

In other research, an indirect axicon-assisted gonioscopy imaging probe with white light illumination that significantly enhances ICA visualization, overcoming resolution limitations in current clinical imaging and providing complementary insights for glaucoma evaluation was developed (Perinchery *et al* 2016). The imaging results from the iridocorneal angle (ICA) region using this optical setup are presented in figure 2.32(a). A more detailed view of the region of interest (ROI) is shown in figure 2.32(b), where digital processing techniques, including edge enhancement and brightness/contrast adjustments, have been applied. Compared to images captured with a Gaussian-illuminated light sheet, this configuration proves more effective for analyzing the aqueous outflow system (AOS), as it

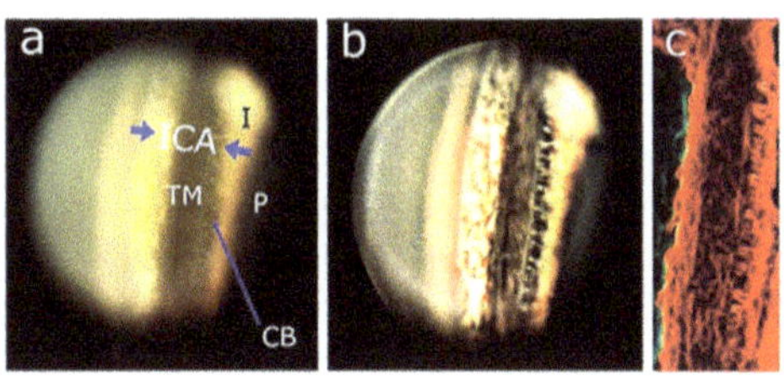

Figure 2.32. (a) Unprocessed image of the trabecular meshwork (TM) region in a porcine eye. (b) Enhanced image of the TM region after processing. (c) Edge-detected image, revealing internal structures within the TM region. [ICA: Iridocorneal angle; P: Pupil; I: Iris; CB: Ciliary body; TM: Trabecular meshwork]. Reproduced with permission from reference (Perinchery *et al* 2016) with permission from Springer-Nature under the license CC BY 4.0.

provides higher axial resolution, enabling clearer visualization of trabecular meshwork (TM) structures. This enhancement results from the thinner light sheet and the Bessel–Gauss beam's ability to penetrate scattering media and reconstruct images. The acquired images maintain sharp focus because the light sheet is thinner than the detection objective's DOF. Additionally, the self-reconstructing nature of the Bessel–Gauss beam improves image contrast at the TM while minimizing scattering and shadowing artifacts (Perinchery *et al* 2016). However, the lateral resolution remains limited by the diffraction constraints of wide-field microscopy, similar to the static light sheet fluorescence imaging.

The findings distinctly reveal a mesh-like network of collagen fibers, with dimensional measurements closely matching those reported in previous histological studies. The image resolution and contrast can be improved much higher as illustrated by adopting such approaches that can improve current clinical imaging techniques and enhance treatment outcomes. However, additional translational research is required prior to its translation in clinical settings.

2.5 Laser tissue interaction—basic optics

This section explores the fundamental principles of light–matter interaction, supported by illustrative examples. It begins with an overview of electromagnetic radiation and the three primary phenomena that occur when light interacts with matter. Additionally, it provides a brief discussion on collimation optics, polarization, and various related optical components that serve as the foundation for laser-tissue interaction and diagnostic medical optics.

2.5.1 Introduction

Light can interact with tissue majorly in four different ways: absorption, transmission, reflection, and fluorescence as shown in figure 2.33.

The electrons in an atom in general are responsible for light interactions and vibrate at its natural frequencies. This interaction between the light and the electron in their frequencies determines whether the incident light is absorbed, reflected or transmitted.

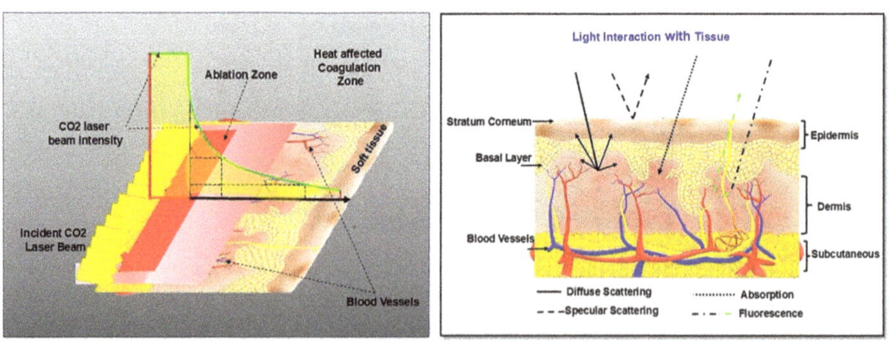

Figure 2.33. Light–tissue interaction illustration.

2.5.2 Reflection, absorption and transmission

Reflection occurs when the frequency of an incoming light wave does not match the natural frequency of the electrons in a material. In opaque objects, where light cannot pass through, the electron vibrations are not transferred inward, as happens during absorption. Instead, the surface electrons vibrate momentarily and then emit the light wave back. The electrons that absorb the light wave, convert it into vibrational energy. This vibrational energy is then transferred to neighbouring atoms, turning into thermal energy.

Transmission operates similarly to reflection but occurs in transparent or semi-transparent materials. In this process, the atoms absorb the incoming light wave, vibrate briefly but with smaller amplitudes (unlike the larger vibrations during absorption), transmit the vibrations through the material, and then re-emit the light wave from the opposite side. Figure 2.34 depicts all the phenomena related to the interaction of light with matter (Jacques 2013).

The absorptivity coefficient is influenced by the same factors that impact reflectivity, leading to variations in its value.

For opaque materials, Reflectivity = 1 − Absorptivity.

For transparent materials, Reflectivity = 1 − (Transmissivity + Absorptivity).

Absorption of optical light in biological tissues occurs when photons are absorbed by molecules, converting the light energy into other forms like heat or another form of energy. This process is wavelength-dependent and influenced by specific chromophores (light-absorbing molecules) present in the tissue. Major factors influencing optical absorption include chromophores in biological tissues such as haemoglobin, melanin, water, lipids, collagen. For example, haemoglobin absorbs strongly in the visible and near-infrared (NIR) regions, particularly at wavelengths around 400–600 nm, and is a major contributor to absorption in blood-rich tissues. The absorption spectrum of tissue varies with the wavelength of light. For example, ultraviolet (UV) light (200–400 nm) is strongly absorbed by DNA, proteins, and

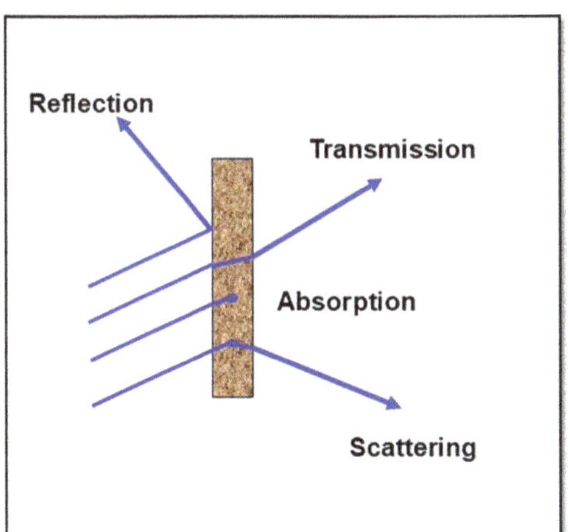

Figure 2.34. Interaction of light with material—major phenomena.

other biomolecules, making it useful for studies of cellular and molecular structures but harmful due to DNA damage. Visible light (400–700 nm is strongly absorbed by haemoglobin and melanin, making it effective for visualizing superficial tissues like skin, but limiting its use for deeper imaging. NIR light (700–1000 nm) is minimally absorbed by water and haemoglobin, making this 'optical window' ideal for deeper tissue imaging with lower absorption and scattering. When the ray enters from one medium to another medium differing in their refractive indices, refraction occurs (Born and Wolf 2020), which is governed by Snell's law given by equation (2.3).

2.5.3 Scattering

Scattering of light in biological tissues occurs due to the interaction of photons with cellular structures, causing light to deviate from its original path as shown in figure 2.35. This phenomenon can be described by the scattering coefficient μ_s, which quantifies the probability of scattering per unit length and size of the particle (Jacques 2013).

Optical scattering in biological tissues can be categorized based on how photons interact with tissue structures. The key types of scattering are listed as follows.

2.5.3.1 Elastic scattering (Rayleigh and Mie scattering)

In elastic scattering, the wavelengths of the light that are scattered remain the same as that of the incident light. This type of scattering can be further divided into:

Rayleigh scattering: It occurs with particles whose size is smaller than the light wavelength. In tissues, Rayleigh scattering primarily affects smaller structures like

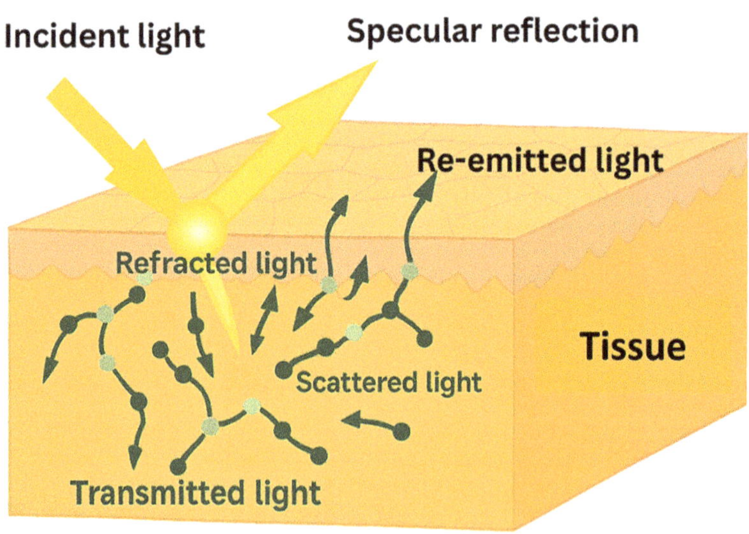

Figure 2.35. Scattering in tissues.

proteins or small organelles. It is more pronounced with shorter wavelengths (e.g., blue light), leading to significant scattering in the visible light range.

Mie scattering: Occurs when the scattering particles are comparable in size to the wavelength of light. In biological tissues, Mie scattering happens with larger structures such as cells and cell nuclei. It affects light across a broad range of wavelengths, including longer wavelengths like near-infrared, and is a dominant form of scattering in tissues.

2.5.3.2 Inelastic scattering (Raman and Brillouin scattering)

In inelastic scattering, the scattered light changes its energy compared to the incident light, often due to interaction with molecular vibrations or other dynamic tissue properties. Two common types of inelastic scattering are:

Raman scattering: Results from the interaction of light with molecular vibrations in the tissue, leading to shifts in the light's wavelength. Raman scattering provides molecular-specific information and is used in techniques like Raman spectroscopy to identify biochemical changes in tissues.

Brillouin scattering: Involves the interaction of light with acoustic phonons (sound waves) in a material. This scattering type is used to study mechanical properties such as tissue elasticity or stiffness at a microscopic level.

2.5.3.3 Quasi-ballistic and diffuse scattering

As photons travel deeper into tissue, they undergo multiple scattering events. Two main forms of scattering occur in this regime:

Ballistic scattering: Refers to light that travels through tissue with minimal scattering, retaining its original trajectory. It is used in imaging techniques like optical coherence tomography (OCT), which captures high-resolution images of superficial layers of tissue.

Diffuse scattering: Happens when light is scattered multiple times, losing its directionality and spreading out as it travels deeper into tissue. Diffuse scattering dominates in highly scattering media like biological tissues and is used in imaging modalities like diffuse optical tomography (DOT) to probe deeper layers.

Each type of optical scattering interacts with light in a unique way, influencing the effectiveness of optical imaging techniques in capturing structural, functional, or molecular details from biological tissues. In biological tissues, scattering is a significant factor that affects the penetration depth and resolution of optical imaging techniques. This limits the depth at which light becomes diffused, making it difficult to visualize structures located deeper within the tissue. The degree of scattering increases with shorter wavelengths of light, such as in the visible range, whereas longer wavelengths, like the near-infrared wavelength band, experience less scattering and allow deeper tissue penetration.

Despite these challenges, optical scattering can be advantageous in certain imaging techniques. For example, techniques like OCT and DOT leverage scattering to provide high-resolution images of superficial tissue layers. However, overcoming the limitations of scattering is essential for advancing deeper tissue imaging, leading to the development of hybrid modalities like photoacoustic imaging, which combines optical and ultrasound techniques to achieve better resolution and penetration.

2.5.4 Fluorescence

Extensive research has been conducted on spectroscopic analysis to identify differences between normal and malignant tissues. The majority of these studies were performed on biopsied tissue samples, with suggestions made for potential *in vivo* applications. UV can excite most fluorophores and spectral information from tissue, especially the surface layers, can help identify disease condition. IR can penetrate deep into the tissue and resulting absorption/Raman shift be studied. Raman/FTIR spectroscopy can in principle provide information on the biochemical makeup of the tissue, but a practical execution for an *in-vivo* diagnostic application is difficult as reflectance signals are weak and the process slow. A trans-illumination scheme with IR for disease diagnosis is more practical for external organs such as breast and limbs, and not suitable for examination of cavities such as the colon.

Fluorescence characteristics of tumours have been studied by spectroscopic methods by various groups with a single excitation wavelength illumination as well as with varying excitation wavelengths (Lu and Fei 2014). An interesting scheme was quoted in a study where a 2D array of optical fibres captured an image and the image was converted into a 1D array at the opposite end of the optical fibre bundle (Lim and Murukeshan 2016). The fluorescent light output from each of the fibres was dispersed by grating optics to generate a two-dimensional spatial-spectral image. Here, the optical fibre size posed an issue (230 μm, compared to CCD resolution of 10 μm). The scheme may be suitable for a very small imaging region with a few fibres in the bundle or (more), and for whole-field imaging (Lu and Fei 2014). In filter-based imaging microscope systems, temporal resolution is constrained by the speed of wavelength switching. This switching occurs in the order of seconds when using a filter wheel, approximately a few tens of milliseconds with a liquid crystal tunable filter (LCTF) and reaches sub-millisecond speeds with an acousto-optical tunable filter (AOTF) (Antony *et al* 2024).

Near-infrared spectroscopy had been employed to identify spectral differences of normal and malignant tissues by various research groups (Amiot *et al* 2008). Challenges were in detecting the weak signals from the soft tissue and masking of signals due to autofluorescence effects. Some papers express caution on reliability of NIR spectroscopy-based diagnosis, as a multitude of factors coupled with weak signals can result in erroneous conclusions (Amiot *et al* 2008).

In a study, laser-induced fluorescence (LIF) with 370 nm excitation was investigated for detection and diagnosis of neoplastic human tissues and it was claimed that with spectral differences analysis, LIF could provide information on the disease state of the tissue in real time, and without any need for tissue biopsy (Niazi *et al* 2022). Advancements in endoscopic imaging have led to the development of specialized systems designed to enhance the detection of abnormal tissue. One such system, as documented in the literature, utilizes an endoscope equipped with distinct charge-coupled devices (CCDs) for red and green channel autofluorescence detection. This technology operates under blue light illumination at a wavelength of 442 nm, allowing for the differentiation of tissue characteristics based on fluorescence emission patterns. Additionally, a separate CCD is incorporated for standard white light imaging, ensuring comprehensive visualization of the examined area. When exposed to 442 nm excitation light, normal tissue exhibits a strong green fluorescence emission, whereas cancerous tissue demonstrates a significantly weaker fluorescence response. This contrast in emission intensity provides a valuable diagnostic cue, aiding clinicians in distinguishing between healthy and malignant regions with greater precision. By integrating autofluorescence with conventional white light imaging, this approach enhances the accuracy of endoscopic examinations and supports early detection efforts in oncology (Takehana *et al* 1995).

Thomas Wang *et al* demonstrated a scheme whereby a single fluorescence emission bandwidth was used for detection (based on intensity differences between normal and adenomatous colon tissue) and instantaneously overlaid, on a white light image to draw the attention of the endoscopist to a suspicious area in the colon (Stoneman *et al* 2024). In a colorectal screening, comparison of ratio of intensities of laser induced autofluorescence emission wavelengths was used to distinguish between normal and adenomatous tissues *in-vivo* using a fibre optic probe and 405 nm laser illumination. Combination of fluorescent and reflective spectroscopy has been suggested to discriminate normal and neoplastic epithelium (Stoneman *et al* 2024). In a multiphoton fluorescence excitation microscopy study, 730 nm laser pulses were used for two-photon excitation of fluorophores with much shorter emission wavelengths, e.g. 450 nm, avoiding the use of UV wavelengths. UV is known to cause photobleaching of fluorophores and phototoxicity in living cells. Also, the use of longer excitation wavelength helped excite fluorophores in deeper tissue layers. fiber evanescent wave spectroscopy (FEWS), differential path length spectroscopy, photoacoustic spectroscopy, low coherence enhanced backscattering (LEBS) and Doppler optical coherence tomography were other methods that have been suggested to identify spectral and structural differences in tissues (Bonin *et al* 2010, Bayer *et al* 2012)

2.5.4.1 Fluorescence imaging

This section reviews key advancements in fluorescence imaging and explores the potential application of fluorescence microscopy for achieving micron or sub-micron spatial resolution at the tissue surface. Fluorescence imaging has garnered substantial attention in biomedicine due to its ability to provide functional and molecular insights into cellular processes. The technique leverages various optical properties of light through photophysical and photochemical mechanisms at the molecular level, enabling the visualization of biomolecular activities in living organisms. This is accomplished when certain molecules absorb light at a particular wavelength and then emit light at a longer wavelength (Amiot *et al* 2008). Fluorescence imaging can be performed using both exogenous and endogenous chromophores. Flavin adenine dinucleotide (FAD) and reduced nicotinamide adenine nucleotide (NADH) are endogenous chromophores and are commonly used for distinguishing between normal, benign, and malignant tissue areas. Disease biomarkers can also be identified by designing molecular probes, which are fluorescent dyes conjugated with targeting molecules. Detection sensitivity can be further enhanced through plasmonic amplification of fluorescent dyes. These methods hold promise for the early diagnosis, prognosis, and staging of various diseases.

2.5.5 Effect of different parameters

The optical properties of the laser influence the surface of the tissue sample where the laser beam strikes. Absorptivity of tissue has the greatest impact on the power needed. Absorptivity itself is affected by factors such as wavelength, surface roughness, temperature, and the sample material's phase.

2.5.5.1 Effect of wavelength

Wavelength is the specific spectral length corresponding to a single cycle of vibration for a photon within a laser beam. The absorptivity of the material depends on the wavelength. Hence, certain lasers are suitable for processing only certain materials. At shorter wavelengths, the energetic photons are absorbed by large numbers of bound electrons and hence the reflectivity falls, and the surface's absorptivity gets increased.

For instance, the extent to which a laser beam's energy is absorbed is influenced by the wavelength of the laser beam and the spectral absorptivity characteristics of the samples or tissues. As a result, this process becomes more efficient, as minimal energy is lost during absorption, maximizing its effectiveness (Jacques 2013).

2.5.5.2 Effect of tissue surface roughness

Surface roughness significantly impacts absorption due to multiple reflections caused by the randomization of the beam. Additionally, 'stimulated absorption' can occur when the beam interferes with sideways-reflected beams. If the roughness is smaller than the wavelength of the beam, these effects are avoided, and the surface appears smooth to the radiation. In such cases, the reflected phase front, generated from Huygens wavelets, will differ from the incident beam, leading to diffuse reflection in various directions. Interestingly, if the reflection from a mirror is perfectly specular, the point where a laser beam strikes the surface should not be visible.

2.6 Maximum permissible exposure (MPE)

Maximum permissible exposure (MPE) refers to the highest level of laser or optical radiation that biological tissue, particularly skin and eyes, can be exposed to without causing harm. MPE values are established to prevent tissue damage like burns, eye injuries, or long-term effects from excessive radiation exposure (Chua *et al* 2017). The MPE varies depending on several factors:

1. **Wavelength of Light:** Different wavelengths interact with tissues differently:
    - Ultraviolet (UV) light (180–400 nm) has stricter MPE limits due to its potential to cause skin burns, DNA damage, and eye injuries like cataracts.

- Visible light (400–700 nm) has more lenient limits, but care is needed with high-intensity exposures as they can still cause retinal burns or skin damage.
- Infrared (IR) light (700–1400 nm) is particularly hazardous to the eye as it can penetrate deeper into the retina without activating the blink reflex, leading to potential thermal damage.

2. **Exposure Duration:** MPE decreases with increasing exposure duration. For example:
   - Short bursts of laser light (pulsed lasers) can have higher MPE compared to continuous-wave lasers, where prolonged exposure can lead to heating and tissue damage.
   - Longer exposure times reduce the MPE to avoid cumulative thermal effects on tissues.
3. **Spot Size or Beam Diameter:** The size of the laser spot on the tissue affects how the energy is distributed:
   - Larger spots distribute energy over a broader area, potentially allowing for a higher MPE.
   - Smaller spots concentrate energy, reducing MPE and increasing the risk of localized damage.
4. **Tissue Type:** Different tissues have varying sensitivities:
   - Eyes are particularly sensitive, and the retina has stricter MPE limits compared to skin due to the potential for irreversible retinal damage.
   - Skin is more resistant but still susceptible to burns or other thermal effects with excessive exposure.
5. **Safety Guidelines:** MPE values are provided by standards such as those from the American National Standards Institute (ANSI) and the International Commission on Non-Ionizing Radiation Protection (ICNIRP). These guidelines help ensure safe use of lasers and optical devices in medical, industrial, and research settings.

Hence, MPE defines the threshold of safe optical exposure to prevent tissue damage, and it is determined by the wavelength, exposure time, beam characteristics, and tissue sensitivity.

2.7 Problems

1. Explain briefly the different phenomena that happen when light interacts with matter using a simple illustrative diagram.
2. How can one obtain circular polarization of light beam?
3. What is Malus' law? Draw the optical configuration to demonstrate Malus' law and mark individual optical components.
4. What is critical angle and how does critical angle play a role in waveguiding in optical fibers?
5. Explain the coherence property associated with laser beams. What are the two different types of coherence?

6. What is a dichroic beam splitter? Explain its working principle and how it differs from a normal cube beam splitter.
7. What are the different laser beam characteristics?
8. How does state of polarization play a role in laser–material/tissue interactions?
9. Why are curved mirrors preferred in laser resonators? Explain with a brief schematic diagram.
10. Explain why a laser beam is much more dangerous to our eyes and skin than lamp light.
11. Explain the maximum permissible exposure (MPE) and explain its dependant parameters.
12. What is the rationale behind using beam expanders to reduce the beam divergence?
13. How can we obtain Bessel–Gauss beams? Explain with schematic diagrams.
14. Explain the relation between NA and the thickness of the virtual light sheet in Bessel–Gauss fluorescence imaging systems.
15. Write a brief note on Bessel–Gauss light beam fluorescence imaging.
16. What are the major differences between Bessel beams and Bessel–Gauss beams?

References

Amiot C L, Xu S, Liang S, Pan L and Zhao J X 2008 Near-infrared fluorescent materials for sensing of biological targets *Sensors (Basel)* **8** 3082–105

Antony M M, Suchand Sandeep C S and Vadakke Matham M 2024 Hyperspectral vision beyond 3D: a review *Opt. Lasers Eng.* **178** 108238

Bayer C L, Luke G P and Emelianov S Y 2012 Photoacoustic imaging for medical diagnostics *Acoust. Today* **8** 15–23

Bonin T, Franke G, Hagen-Eggert M, Koch P and Huttmann G 2010 *In vivo* Fourier-domain full-field OCT of the human retina with 1.5 million A-lines/s *Opt. Lett.* **35** 3432–4

Born M and Wolf E 2020 *Principles of Optics* 7th edn (Cambridge: Cambridge University Press)

Chua C K, Murukeshan V M and Kim Y-J 2017 *Lasers in 3D Printing and Manufacturing* (Singapore: World Scientific)

Herman R and Wiggins T 1991 Production and uses of diffractionless beams *JOSA* A **8** 932–42

Hong J X J 2018 Investigations into high resolution imaging of the aqueous outflow system and cornea *Doctoral thesis* (Singapore: Nanyang Technological University)

Hong X J J S, Sandeep C S, Shinoj V K, Aung T, Barathi V A, Baskaran M and Murukeshan V M 2020 Noninvasive and noncontact sequential imaging of the iridocorneal angle and the cornea of the eye *Transl. Vis. Sci. Technol.* **9** 1

Hong X J J, Shinoj V K, Murukeshan V M, Baskaran M and Aung T 2017 Imaging of trabecular meshwork using Bessel–Gauss light sheet with fluorescence *Laser Phys. Lett.* **14** 035602

Jacques S L 2013 Optical properties of biological tissues: a review *Phys. Med. Biol.* **58** R37–61

Lim H T and Murukeshan V M 2016 A four-dimensional snapshot hyperspectral video-endoscope for bio-imaging applications *Sci. Rep.* **6** 24044

Lu G and Fei B 2014 Medical hyperspectral imaging: a review *J. Biomed. Opt.* **19** 10901

Niazi A *et al* 2022 Discrimination of normal and cancerous human skin tissues based on laser-induced spectral shift fluorescence microscopy *Sci. Rep.* **12** 20927

Perinchery S M, Haridas A, Shinde A, Buchnev O and Murukeshan V M 2019 Breaking diffraction limit of far-field imaging via structured illumination Bessel beam microscope (SIBM) *Opt. Express* **27** 6068–82

Perinchery S M, Shinde A, Fu C Y, Jeesmond Hong X J, Baskaran M, Aung T and Murukeshan V M 2016 High resolution iridocorneal angle imaging system by axicon lens assisted gonioscopy *Sci. Rep.* **6** 30844

Shinoj V K, Hong X J J, Murukeshan V M, Baskaran M and Aung T 2016 Progress in anterior chamber angle imaging for glaucoma risk prediction—a review on clinical equipment, practice and research *Med. Eng. Phys.* **38** 1383–91

Silfvast W T 2012 *Laser Fundamentals* 2nd edn (Cambridge: Cambridge University Press.)

Stoneman M R, McCoy V E, Gee C T, Bober K M M and Raicu V 2024 Two-photon excitation fluorescence microspectroscopy protocols for examining fluorophores in fossil plants *Commun. Biol.* **7** 53

Suchand Sandeep C S, Chan Lwin N, Liu Y-C, Barathi V A, Aung T, Baskaran M and Murukeshan V M 2022 High-resolution, non-contact, cellular level imaging of the cornea of the eye *in vivo Opt. Laser Technol.* **150** 107922

Suchand Sandeep C S, Sarangapani S, Hong X J J, Aung T, Baskaran M and Murukeshan V M 2019 Optical sectioning and high resolution visualization of trabecular meshwork using Bessel beam assisted light sheet fluorescence microscopy *J. Biophotonics* **12** e201900048

Takehana S, Kaneko M and Mizuno H 1995 Endoscopic diagnostic system using autofluorescence. *Diagn. Ther. Endosc.* **5** 59–63

IOP Publishing

Diagnostic Biomedical Optics

Fundamentals and applications

Murukeshan Vadakke Matham, C S Suchand Sandeep, Maria Merin Antony, Manojit Pramanik and Santhosh Chidangil

Chapter 3

Diagnostic biomedical imaging—fundamentals

C S Suchand Sandeep and Murukeshan Vadakke Matham

This chapter discusses major diagnostic biomedical optical imaging modalities and the principles behind them. The chapter starts with a discussion of wide-field imaging systems followed by confocal imaging systems. It then introduces photoacoustic microscopy, optical coherence tomography and spectroscopic imaging systems. It concludes with a discussion on structured illumination microscopy and laser speckle imaging. The advantages, limitations and biomedical applications of each of these imaging modalities are detailed.

3.1 Wide-field imaging systems

Wide-field optical microscopy is one of the most common tools used for biomedical imaging applications. In this technique, the sample as a whole is illuminated by a light source either from the bottom (usually in the upright microscope configuration) or from the top (usually in the inverted microscope configuration). The selection of the microscope configuration generally depends on the sample type to be investigated. Samples that are fixed onto microscope slides or that are in the form of a thin film are usually examined using the upright microscope. For samples dispersed in a liquid, often the inverted microscope configuration is preferred as the samples are easier to see from the bottom side in these suspensions. Several configurations of wide-field microscopy exist, the most prevalent ones are discussed in the following sections.

3.1.1 Brightfield microscopy

In this type of microscope, the whole sample is lit by a bright light source. The microscope objective collects the transmitted or reflected light from the sample and images it on to the detector. The samples usually absorb or reflect the incident light, thereby altering the amplitude of the light passing through them. Since the material

doi:10.1088/978-0-7503-2364-2ch3

and thickness of the sample are space variant, these variations accordingly create a space-variant change in the amplitude of the light passing through the sample. This amplitude variation is captured by the detector to create the image of the object. Such objects that absorb light to create an amplitude variation in the light passing through them are called *amplitude objects*. The background or the space between the objects in the sample does not absorb light and appear bright in the images recorded. This is the reason for this technique being named brightfield imaging. The advantages of this approach are that it requires minimal sample preparation, and live cells can be quickly examined (Swedlow and Platani 2002). However, there are several disadvantages, such as the lack of contrast (especially for transparent samples), limited resolution, higher chances of photodamage (damage caused to the samples by high amount of light irradiation), etc. Most often, contrast agents (like stains or dyes) are used for improving the contrast in cellular imaging as most cell samples are quite transparent to visible light (Shen and Zhang 2023). The typical imaging scheme used for brightfield imaging is shown in figure 3.1.

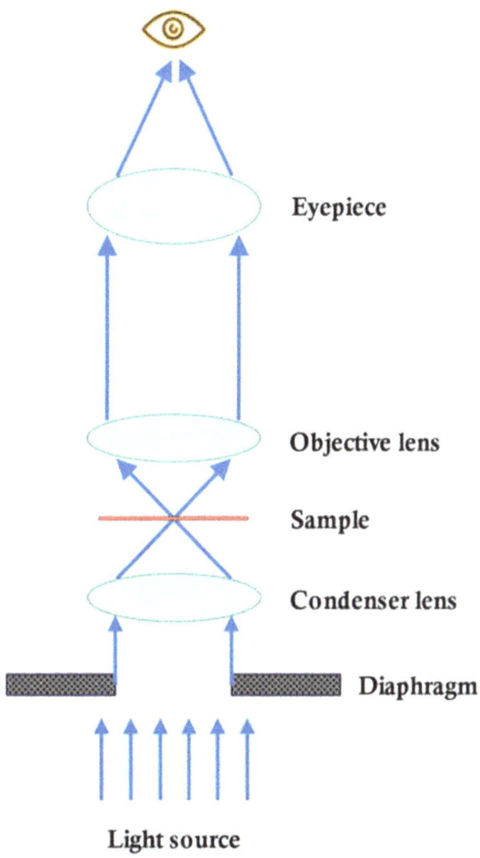

Figure 3.1. Illustration showing the typical imaging scheme used for brightfield imaging.

3.1.2 Darkfield microscopy

Darkfield microscopy makes use of oblique illumination for contrast enhancement in samples that are difficult to image under brightfield imaging. In darkfield microscopy, the centre of the illumination is blocked by an opaque stop placed before the condenser lens, or the illumination is done using a ring-shaped beam (also often referred to as hollow cone illumination). This helps in preventing light transmitted directly through the sample from reaching the objective lens. Only the light that is scattered, diffracted, or reflected by the structures in the sample will reach the microscope. The sample thus appears bright against a dark background. This helps in getting a better contrast than in brightfield imaging. Figure 3.2 illustrates a typical darkfield microscopy configuration.

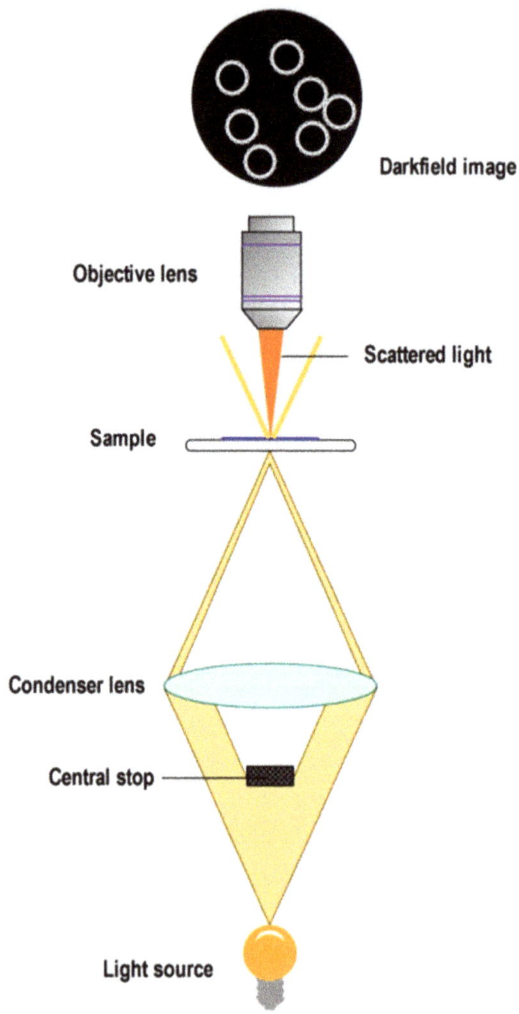

Figure 3.2. Schematic of a typical darkfield microscopy configuration.

Though darkfield microscopy provides better contrast, it comes at the cost of the brightness of the images. The amount of light reaching the objective is quite low compared to brightfield imaging scheme, and often to compensate, higher illumination intensities are used. This may lead to photodamage in the sample. In addition, careful sample preparation is also necessary, as any debris or dust on the sample surface will also scatter the light and appear bright in the images recorded.

3.1.3 Phase contrast microscopy

Transparent objects, unlike the amplitude objects, do not cause any amplitude variations in the light transmitted through them. However, they impart a delay to the light wave passing through them, causing a phase delay (the higher the refractive index, the higher the phase delay) in the light wave. Thus, these objects are called *phase objects*. Our eyes, as well as commonly used photodetectors such as CCD cameras, are not capable of detecting the phase changes and hence it is difficult to image transparent samples in wide-field imaging. In phase contrast microscopy, these phase changes are converted to amplitude changes, thereby providing a way to image transparent objects with discernible contrast (James and Tanke 1991). This is achieved by introducing a suitable phase change in the direct light transmitted through the sample, so that it destructively interferes with the scattered/diffracted light and forms an amplitude image at the detector. One of the phase contrast microscopy schemes is shown in figure 3.3.

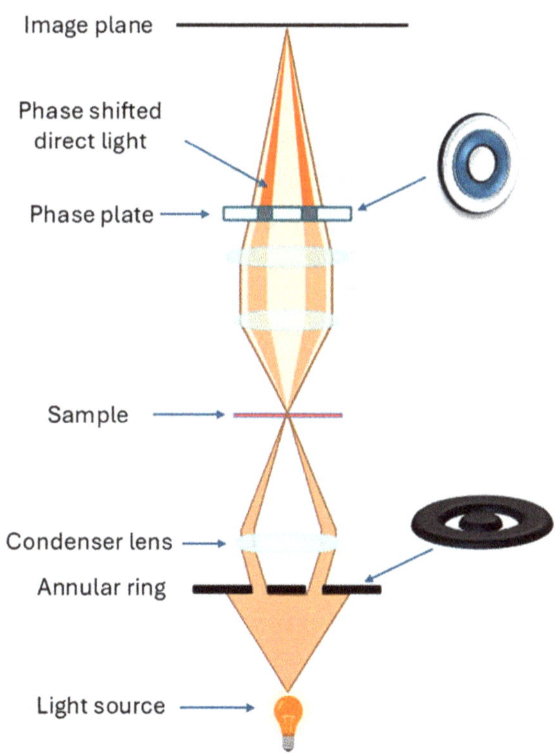

Figure 3.3. Illustration showing phase contrast microscopy configuration.

In this scheme, the light is first filtered by a specialized annulus ring (shown on the right side of figure 3.3), which is then focussed by the condenser lens to create a ring-like illumination at the sample. This annulus illumination passing through the sample is either scattered, diffracted, or delayed in phase by the sample structures or propagates undeviating. All the light emanating from the sample is collected by the microscope objective and the undeviating light is allowed pass through a phase plate placed in the rear focal plane (shown on the top right side of figure 3.3). The final phase-contrast image is formed at the intermediate image plane after the phase plate due to interference, which can be observed through the eyepiece or recorded by a camera (James and Tanke 1991).

3.1.4 Differential interference contrast (DIC) microscopy

Differential interference contrast is another contrast-enhancing technique used for *phase objects*. It makes use of the polarization property of light to convert phase information into contrast (Ruzin 2024). The light from the source is linearly polarized using a polarizer and subsequently split into two orthogonally polarized components that are slightly shifted in space (sheared) with the help of a Wollaston prism. The Wollaston prism is made by cementing two birefringent wedges that have optical axes perpendicular to each other. Birefringent materials are optically anisotropic materials whose refractive index is dependent of the polarization and direction of propagation (Ruzin 2024). Two perpendicularly polarized beams result after passing through the Wollaston prism, which is allowed to pass through the sample. The beams are recombined using a second Wollaston prism placed after the sample. Because of the shear, the beams experience slightly different path lengths while traversing through the sample. This results in an elliptically polarized beam to be generated when they recombine, which is analysed using another polarizer (known as analyzer). The phase objects thus can be converted to an amplitude object. Usually, the DIC microscope generates images that have relief-like appearance. This pseudo 3D-like appearance however is not directly related to the sample geometric thickness but is rather related to the optical pathlength through the sample (Ruzin 2024). Figure 3.4 shows the schematic of a DIC microscope.

Though wide-field imaging is a simpler method, and several contrast enhancing techniques exist, they still suffer from limited resolution, especially in the axial direction (James and Tanke 1991). The lateral resolution (also often denoted as spatial resolution) of an imaging system refers to the resolution achievable in-plane (*xy*-plane), while the axial resolution refers to the resolution achievable in the axial direction (*z*-direction). A major cause for limited axial resolution is the presence of background illumination from out-of-focus planes (Jerome and Price 2018). The following sections discuss several advanced biomedical optical imaging techniques that overcome some of the limitations of wide-field imaging systems.

3.2 Confocal imaging systems

Confocal microscopy is an advanced imaging technique that can provide contrast enhancement as well as improved resolution (especially axial resolution). Confocal microscopy allows high resolution optical sectioning of samples, thereby enabling the creation of 3D volumetric images of the samples (St. Croix *et al* 2005, Jerome

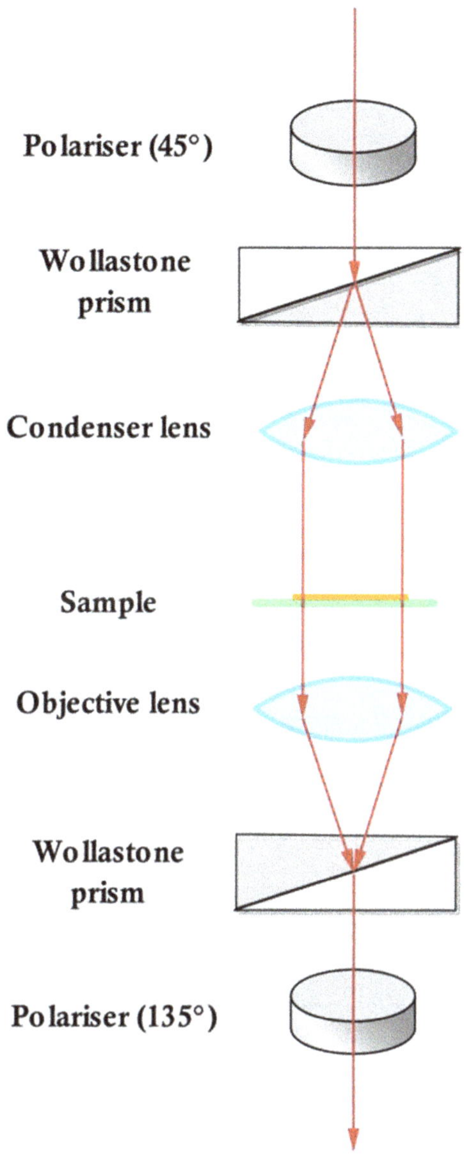

Figure 3.4. A schematic of the differential interference contrast (DIC) microscope.

and Price 2018). The resolution enhancement is achieved by the selective removal of background light with the use of appropriately placed pinholes. Figure 3.5 shows the images of hepatic stellate cells recorded using a confocal imaging system as well as a wide-field imaging system (St. Croix *et al* 2005). The higher resolution and contrast as well as the background rejection from out-of-focus planes is evident in the confocal microscope image. It should, however, be noted that the imaging time required for obtaining the confocal image is much higher than wide-field imaging (~1 h for confocal versus 1 s for wide-field). It should be noted that the images

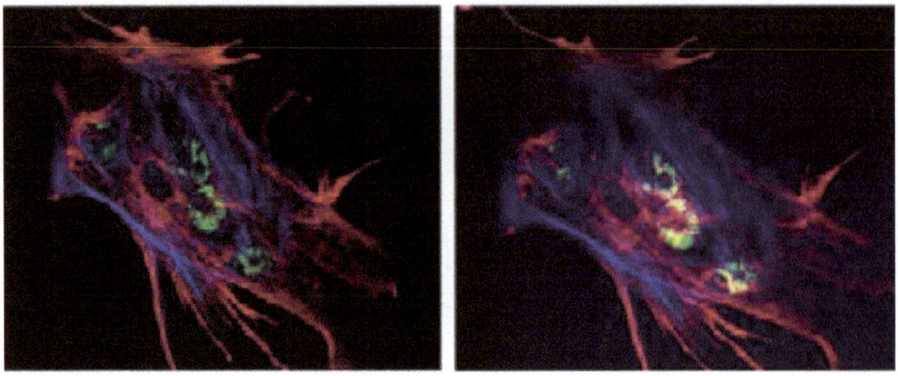

Figure 3.5. Comparison of confocal and wide-field imaging. Confocal microscope image of hepatic stellate cells is shown on the left (3D reconstruction from optically sectioned images recorded) and the wide-field image of the same is presented in the image on the right side. Reproduced from St. Croix *et al* (2005).

shown in figure 3.5 are recorded in the fluorescence imaging modality (discussed later in section 3.5.1).

Several implementations of the confocal microscope exist, and the following sections detail the most prominent ones.

3.2.1 Point scanning confocal systems

The most common version of the confocal microscope is the spot-scanning confocal system, which reconstructs the image of a sample surface from data collected by point-by-point scanning of the sample (Jerome and Price 2018). Typically, a laser is often used as the illumination source to enhance the contrast further. Such a system is commonly known as the laser scanning confocal microscope (LSCM). The strategically placed pinholes in the illumination and imaging arms help to effectively block the light rays emanating from the background (out of focus light), thereby enhancing the resolution. The basic schematic of the point scanning confocal microscope is given in figure 3.6.

In wide-field microscopy, the depth of focus of the system can be a few tens of micrometers (which depends mainly on the objective lens's numerical aperture (NA)). The out of focus light from the regions above and below the focus plane also reaches the objective lens and gets mixed with the focus plane image. This reduces the image contrast significantly and it also has a marked effect on the achievable resolution (Jerome and Price 2018). In LSCM, by virtue of the confocal pinhole, such out of focus light are prevented from reaching the detector enabling high contrast imaging as well as providing good axial resolution. LSCM enables imaging very thin layers (typically a few micrometers) of a thick sample. This is often known as optical sectioning, enabled by the high axial resolution achievable in confocal microscopy (St. Croix *et al* 2005, Jerome and Price 2018). The thickness of such an optical slice can be brought down to about 500 nm using visible light excitation using LSCM. By recording several such optical slices along the thickness of the sample and then stacking them together, the high-resolution 3D volumetric imaging of the sample can be attained. The confocal scanning principle can also be used for improving the

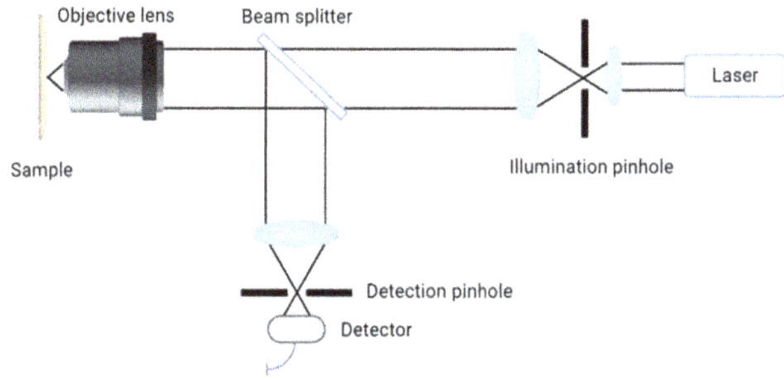

Figure 3.6. Illustration showing the principle behind a point scanning confocal microscope.

resolutions in fluorescence imaging (discussed in section 3.5.1). A major disadvantage of LSCM is the long image acquisition times required due to the raster and *z*-direction scans required to achieve volumetric 3D images (Jerome and Price 2018). To improve the speed of confocal microscopy, novel techniques were developed. Significant advancement was brought about by replacing the single pinhole by a set of spinning disk pinholes (St. Croix *et al* 2005, Jerome and Price 2018). This technique is known as the spinning disk confocal technique and is described in detail in the following section.

3.2.2 Spinning disk confocal systems

In this technique, an opaque disk with multiple pinholes, usually arranged in a helical fashion (also known as a Nipkow disk), is used instead of the single pinhole in the conventional confocal system. When the disk is spun rapidly (typically at a few thousand rpm), the pinholes scan the entire sample. This generates confocal images of the whole sample that can be directly viewed by the eye or by a camera, unlike the point-by-point scanning employed in the traditional confocal microscope (St. Croix *et al* 2005). This offers imaging speeds much faster than the raster point by point scanning scheme. The improved imaging speed in the spinning disk confocal system however comes at the cost of a lowered resolution and contrast. An illustration of the spinning disk confocal system is depicted in figure 3.7.

3.2.3 Chromatic confocal systems

In chromatic confocal systems, white light is used to illuminate the sample surface. The white light, when passed through a specialized sphero-chromatic lens, a continuum of monochromatic components spread along the *z*-axis can be obtained, which can be thought of as a colour-coded *z*-axis. The detection arm contains a filtering pinhole as in the traditional confocal microscopy system. Upon placing an object in this colour-coded field, only the wavelength that gets focused on to the sample surface is reflected, which is then captured by a spectrometer through the confocal pin hole. The spectrometer analyses the highest wavelength component in the spectrum, which is then converted to the corresponding height (*z*-value). Such systems have the advantage of higher speed compared to conventional raster scanning confocal systems and does not require complex algorithms for image reproduction. A schematic illustration of the chromatic confocal system is given in figure 3.8.

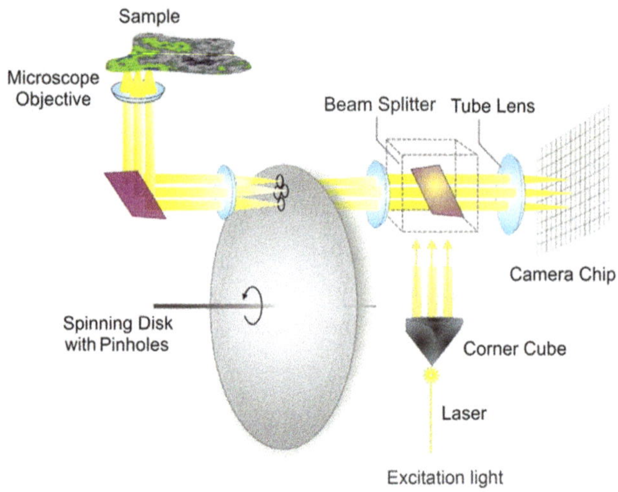

Figure 3.7. Illustration showing the working principle of a spinning disk confocal system. Adapted from Corydon *et al* (2016). CC BY 4.0.

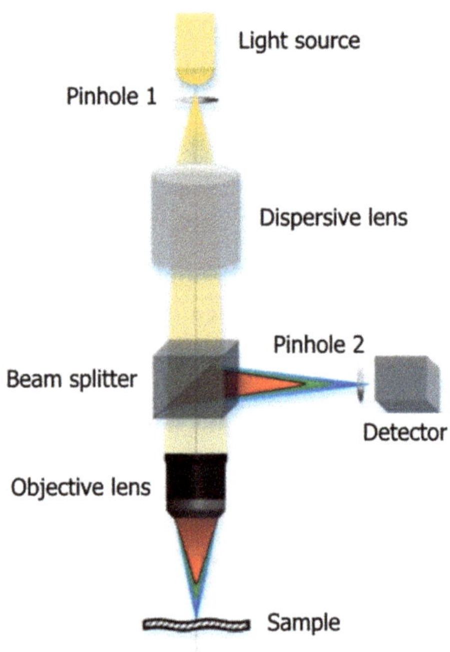

Figure 3.8. Illustration showing the working principle of a chromatic confocal system. Adapted from Yu *et al* (2022). CC BY 4.0.

3.3 Photo-acoustic imaging (PAI)

Photo acoustic imaging (PAI) is a hybrid imaging technique achieved by combining optics and ultrasonics. PAI can provide anatomical as well as functional characterizations of the biological specimen under investigation (Wang 2017). The basis of PAI is the photoacoustic effect, in which the optical excitation of a sample produces acoustic waves (PAI is also often known as opto-acoustic imaging). When the samples absorb energy from the optical excitation beam, it causes thermo-elastic expansion, generation of heat and ultrasonic waves (Beard 2011, Wang 2017). The generated ultrasonic signals can be detected by appropriate transducers and can aid in the anatomical image reconstruction (see figure 3.9). Since the detection is based on ultrasound, PAI offers deeper imaging capabilities compared to other optical imaging modalities. Since photo acoustic signals generated are dependent on the tissue properties and the surrounding environment, it can also be utilized for the quantification of physiological parameters, such as blood flow and oxygen saturation (Beard 2011). By employing time-resolved photoacoustic measurements, the depth information can be attained, which combined with raster scanning of the beam enables 3D volumetric imaging capabilities.

PAI enables deeper imaging capabilities for biomedical investigations. However the disadvantages of conventional PAI include lower contrast, and lower resolution compared to optical imaging (Beard 2011). Recently, by using focused laser beam geometries for excitation, optical resolution photoacoustic modalities have been developed (Beard 2011). In this case, the optical confinement is utilized for localization to improve the spatial resolution. In order to improve the specificity and sensitivity of PAI, exogenous contrast agents can be utilized (Raveendran *et al* 2018, Farooq *et al* 2022).

3.4 Optical coherence tomography (OCT)

Optical coherence tomography (OCT) is a non-invasive optical imaging modality that can be used to image internal structures of biological samples in real time. OCT utilizes low-coherence interferometry to achieve depth resolution and can be applied

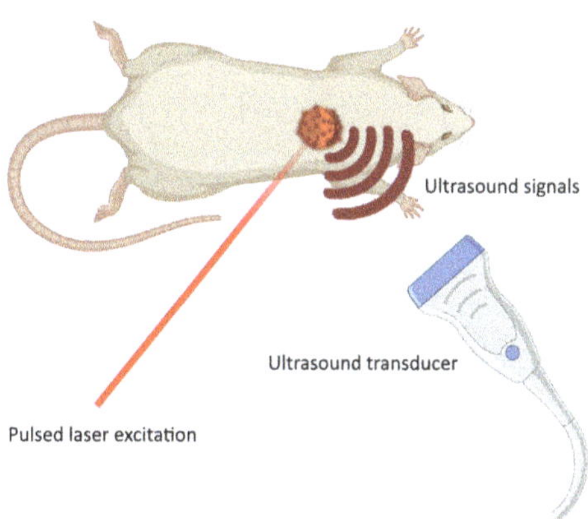

Figure 3.9. A cartoon showing the concept of photoacoustic imaging.

to semi-transparent samples without the need for any sample preparation (Huang *et al* 1991, Drexler and Fujimoto 2015, Girach and Sergott 2016). OCT utilizes the wave nature and interference property of light for generating depth information. The interaction of two coherent beams will lead to constructive or destructive interference depending on the phase difference between the two interacting beams (Born *et al* 1999). In low-coherence interferometry, an interference pattern is generated only when the pathlength difference between two interfering beams are very small. A typical OCT configuration utilizes a Michelson interferometer geometry (Huang *et al* 1991, Drexler and Fujimoto 2015). The light from a low coherence source (such as a super luminescent diode) is split into two arms and the sample sits on one arm of the interferometer. The other arm consists of a reference mirror in a scanning optical delay line. When the optical paths of the two arms are the same, interference patterns are generated and are recorded by the detector. Thus, by scanning the reference mirror in the delay line, the depth-reflectivity profile of the sample can be obtained (Aumann *et al* 2019, Cabaleiro *et al* 2019). The two-dimensional enface profile of the sample can be obtained by scanning the sample laterally (Aumann *et al* 2019). Figure 3.10 shows a simplified schematic of an OCT, depicting the basic principle utilized.

Since OCT commonly utilizes NIR and IR radiation, it can penetrate deeper into the biological tissues and can be described as the optical counterpart to ultrasound imaging. The technique being non-contact, the advantages are several, especially in clinical cases where biopsies can be risky. Another advantage is rapid image acquisition, enabling real-time imaging capabilities and dynamic sample assessment. One of the inherent advantages of OCT is that the lateral and axial resolutions are completely decoupled (Aumann *et al* 2019). Since OCT makes use of low-coherence interference to image the depth field, the axial resolution is defined as the full width at half maximum (FWHM) of the coherence length of the source (l_c). Thus, the axial resolution for a light source having Gaussian spectral distribution can be written as (Aumann *et al* 2019):

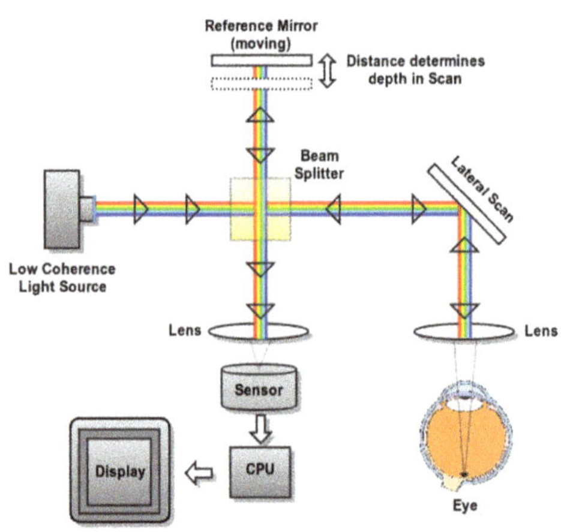

Figure 3.10. A schematic of an optical coherence tomography system. Reproduced from Cabaleiro *et al* (2019). CC BY4.0.

$$\Delta z \sim l_c = \frac{2\ln 2}{\pi} \times \frac{c}{\Delta\nu} = \frac{2\ln 2}{\pi} \times \frac{\lambda^2}{\Delta\lambda} \approx 0.44\frac{\lambda^2}{\Delta\lambda} \tag{3.1}$$

where λ is the central wavelength and $\Delta\lambda$ is the spectral width of the source. With recent developments in mode-locked lasers and photonic crystal fiber lasers, sub-micron coherence lengths and similar axial resolutions have been achieved. The lateral resolution in the OCT system is similar to other microscopy schemes and is governed by the spot size of the probe beam. The lateral resolution in this case is defined as the beam diameter at half maximum, given by (Aumann *et al* 2019):

$$\Delta x = \sqrt{2\ln 2} \times \omega_0 = \sqrt{2\ln 2} \times \frac{2\lambda}{\pi} \times \frac{f}{n \times d} = \sqrt{2\ln 2} \times \frac{\lambda}{\pi \times NA} \tag{3.2}$$

OCT is one of the gold standards used in ophthalmology and had also found applications in other biomedical areas such as dermatology, dentistry and cardiology. Though the initial schemes for OCT involved scanning the beam in a raster fashion to generate enface images (2D plane images) and scanning the reference mirror to generate depth profiles, advanced modalities of OCT have made it possible to achieve full field imaging without the need for scanning for generating enface images. Some of these advancements in OCT are discussed in the following subsections.

3.4.1 Fourier domain-OCT (FD-OCT)

Fourier domain-OCT (also known as frequency domain OCT) works in the same principle based on a Michelson interferometer. However, in this case the mirror is stationary and the spectral information in the interference pattern is spatially spread using a dispersive element (such as a grating), which is then recorded by an array detector. The information about the sample's depth profile is encoded in the spectrum of the interference signal and a Fourier transform of the spectral profile gives information equivalent to that obtained with a moving reference mirror (Drexler and Fujimoto 2015, Aumann *et al* 2019). FD-OCT thus avoids the need for scanning the reference mirror and improves the imaging speed. The implementation of FD-OCT using a broadband source and dispersive elements is called spectral domain-OCT (SD-OCT) (Drexler and Fujimoto 2015). Another way to achieve spectral information in OCT is by sweeping the wavelength of the excitation light over a range of wavelengths and recording the interferograms corresponding to each wavelength. By correlating the sweep parameters with the time coded detector data, the spectral information can be attained. This implementation of FD-OCT is known as swept-source-OCT (SS-OCT). Figure 3.11 illustrates the two implementations of FD OCT (Drexler *et al* 2014).

3.4.2 Full field-OCT (FF-OCT)

In a conventional OCT scheme, the beam is raster scanned to create the enface images. Reconstruction of the volumetric 3D data in this method needs longer times to complete and can also cause motion artifacts in the images due to the mechanical

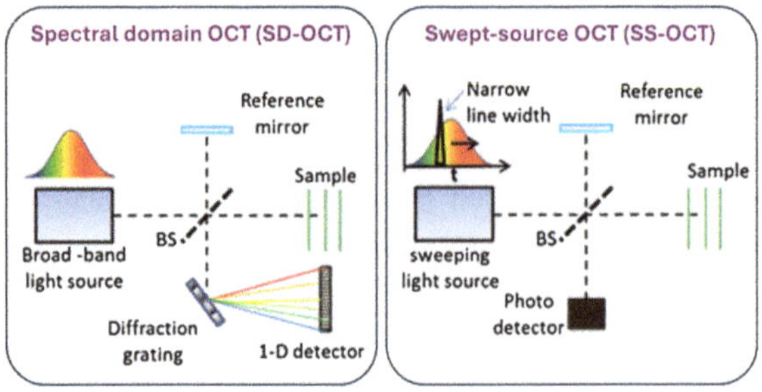

Figure 3.11. Illustrations showing SD-OCT and SS-OCT implementations under the domain of FD-OCT. Adapted from Drexler *et al* (2014) under CC BY 3.0.

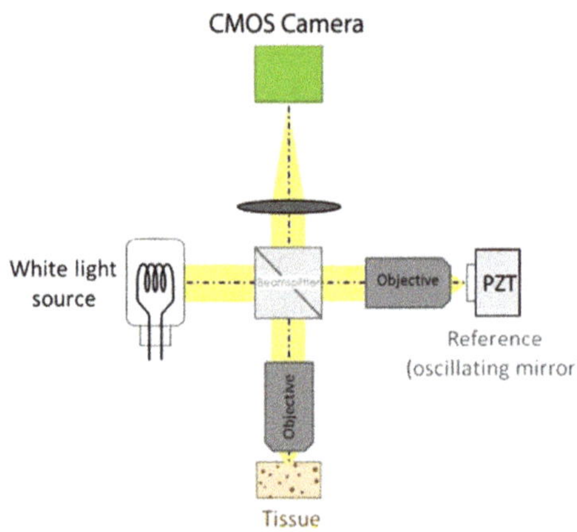

Figure 3.12. Basic scheme of a full field optical coherence tomography (FF-OCT) system. Reproduced from Leroux *et al* (2015) under CC BY 3.0.

motion involved (Dubois and Boccara 2006). Full field-OCT systems make use of 2D array detectors such as a CCD detector to record the enface data without the need for raster scanning (Grieve *et al* 2005). Typically, in this scheme, identical, high NA microscope objectives are used on both arms of the Michelson configuration. The CCD camera can then record the enface images directly. The scanning of the reference arm mirror is still required for recording the depth profile. And due to the low frame rates of the 2D array detectors, the depth scan speeds are lower compared to the point scanning scheme. Recent advancements in this field have enabled the development of a single shot FF-OCT scheme, which provides the feasibility for dynamic imaging of biological specimens (Hrebesh *et al* 2009). The basic setup used in FF-OCT is given in figure 3.12.

3.5 Spectroscopic imaging systems

Spectroscopy refers to the investigation of wavelength dependent properties of a sample. Spectroscopic measurement is one of the most commonly used analysis tools in biomedical imaging, especially for diagnosis purposes. Photons are either absorbed, reflected, or scattered during the interaction of light with a biological sample. In some samples, the absorption of light may lead to fluorescence emission, in which photons of longer wavelength are emitted (Mondal and Diaspro 2014). Several spectroscopic imaging techniques, namely fluorescence imaging, spectral imaging, Raman imaging, etc., make use of the wavelength dependent properties of the samples for analysing them. In this section some of the prominent spectroscopic imaging techniques used in biomedical diagnosis are introduced.

3.5.1 Fluorescence imaging

Fluorescence refers to the emission of a photon with a different wavelength by a material after it absorbs a photon. Typically, the emitted photon has a wavelength that is higher than the absorbed photon. This provides a means of spectrally separating the emitted light from excitation light, providing high contrast images. Fluorescence microscopy's potential was not fully realized until the early 1940s, when fluorescent labelling techniques for antibodies were introduced, despite the fact that fluorescence had been reported as early as in the 1850s (Coons *et al* 1942). Today, fluorescence microscopy is an essential tool in biology owing to its specificity both in endogenous autofluorescence and in selective exogenous fluorescence labelling (Mondal and Diaspro 2014).

In fluorescence emission process, initially the fluorescent molecule absorbs a photon, which has a photon energy equal to the molecular energy gap. In the case of such molecular systems, the excitation photon wavelength can be calculated as

$$\Delta E_g = h\nu = hc/\lambda \tag{3.3}$$

$$\lambda = hc/\Delta E_g \tag{3.4}$$

where λ is the wavelength, h is the Plank's constant, c is the speed of light, and ΔE_g is the molecular energy gap. Following the excitation of the fluorescent molecule, the molecules decay to an intermediate state and then relax back to the ground level by emitting a photon of longer wavelength compared to the excitation wavelength. The difference in excitation and emission wavelengths varies from a few tens to a few hundreds of nanometres. In certain samples, the biological components in the sample itself may possess fluorescence property (known as endogenous fluorescence) (Chorvat and Chorvatova 2009). In other samples, the components could be made fluorescing by selectively staining or labelling them with external fluorescent probes (known as exogenous fluorescence).

The schematic of a basic fluorescence microscope is shown in figure 3.13. The scheme used is very similar to bright field imaging, the main difference being the inclusion of excitation and emission filters. These filters allow only a certain range of wavelengths to pass through them, thereby providing a means of effectively separating the excitation beam from the fluorescent emission. The excitation filter is used to choose the right wavelength for exciting the fluorophore (as given in equation (3.4)), while the emission filter is used to selectively allow only the

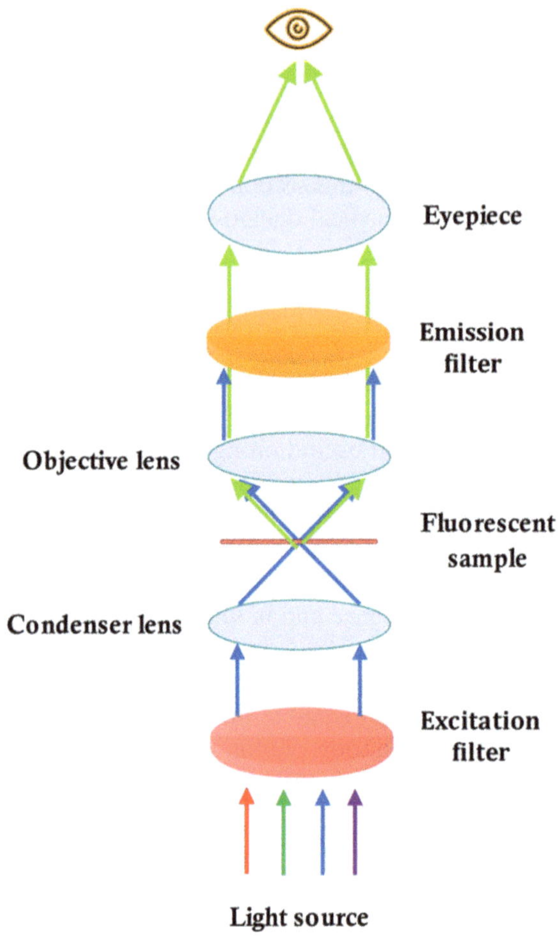

Figure 3.13. Schematic showing a fluorescence imaging system. The excitation and emission filters can be varied to match the specific fluorophore in the system offering versatility and multiprobe imaging capability.

fluorescence emission to pass through. This helps in preventing the background illumination from reaching the detector, thereby increasing the image contrast.

Selective staining of multiple components in a biological sample with different fluorescent probes or utilizing the autofluorescence property of different components in the biological sample, sequential imaging of these components can be achieved, and the images can be combined to get stunning images detailing the structure of the biological sample. Figure 3.14 shows a multi fluorophore probe tagged fluorescent microscope image of bovine endothelial cells. Microtubules are stained green, nuclei are stained blue, and actin filaments are stained red.

It should be noted that there are also some disadvantages associated with fluorescence imaging. The use of external fluorescent probes may cause undesirable effects in the sample under investigation. Some of the fluorescent dyes are toxic/phototoxic in nature and may not be suitable for live sample imaging (Jensen 2012).

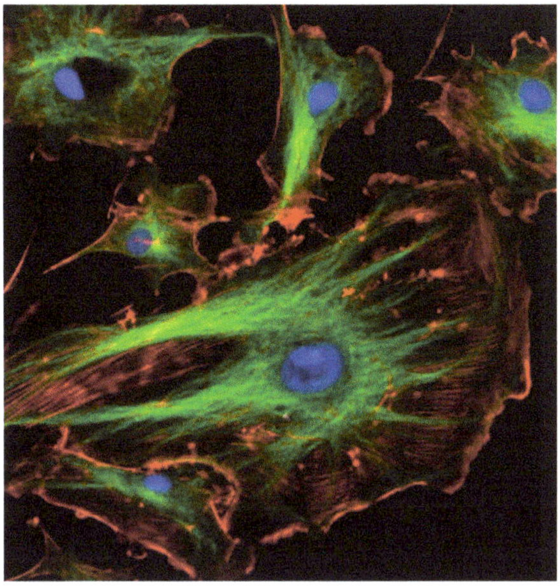

Figure 3.14. Fluorescence images of bovine endothelial cells tagged with multiple fluorescent probes. The actin filaments are stained red using Texas Red X-Phalloidin, microtubles are stained green by the antibody Bodipy FL goat anti-mouse IgG, and the nuclei are marked blue with the dye, DAPI (4′,6-diamidino-2-phenylindole). Reproduced from Rasband (1997–2018). Image stated to be in the public domain.

In addition, the dyes may not cover the whole sample and this could result in incomplete shape information being recorded (Jensen 2012). The information recorded in fluorescence imaging is the fluorescence activity and this may not correlate directly with the structural changes in the sample. Several variants of fluorescence microscopy exist, such as bright field fluorescence microscopy, epi-illumination fluorescence microscopy, confocal fluorescence microscopy, and fluorescence lifetime imaging microscopy (FLIM). FLIM is an important tool used in biomedical imaging (Suhling *et al* 2015). Most fluorescence imaging methods record the integrated intensity of the fluorescent light over time, generating images that represent fluorescence intensity distribution in the sample. In contrast, in FLIM technique, the evolution of the fluorescent signal intensity with time is recorded with high temporal resolution (nanoseconds or picoseconds resolution). The average duration a fluorophore molecule spends in the excited state before returning to the ground state is referred to as the fluorescence lifetime of that molecule. This is greatly influenced by the local microenvironment and hence provides important information about the sample. Though the implementation is intricate, FLIM can offer unique insights into biological activities, dynamics, and interactions in cells, with high spatial and temporal resolution (Berezin and Achilefu 2010, Suhling *et al* 2015).

3.5.2 Raman imaging

Another important spectroscopy tool that offers highly specific microscopic information is Raman imaging. As the name suggests, this technique uses the Raman effect to create microscopic images of the bio sample with molecular specificity. The Raman effect arises from the inelastic scattering of photons by atoms or molecules. Most of the photons that fall on an atom or molecule gets elastically scattered (known as Rayleigh scattering, there is no change in the scattered photon energy). However, a small fraction of the photons gets scattered inelastically, and in the process the scattered photon energy gets altered (Zoubir 2012). This shift in the energy of the scattered photons can be directly correlated to the vibrational energy states of the scattering atom or molecule and thus acts as a fingerprint of the atom or molecule. Raman scattering is broadly categorized into Stokes and anti-Stokes scattering. In Stokes scattering, the Raman lines are observed at energies lower than the exciting photon energy, while in anti-Stokes scattering, the Raman lines have energies higher than the exciting photon energy (Zoubir 2012). The excitation is often carried out with a monochromatic laser beam, and in principle, any excitation wavelength can be used as only the shift in the scattered photon wavelength is important in Raman spectroscopy/imaging (Turrell and Corset 1996). In a typical Raman imaging microscope, the excitation laser is focussed onto the sample, and the scattered light is collected by the microscope objective. The reflected and Rayleigh scattered light are filtered out using a notch filter at the laser wavelength or using a long pass filter with a cutoff wavelength slightly higher than the excitation laser wavelength. The Raman signals collected are analyzed using a highly sensitive spectrometer and sample scanning schemes are utilized to generate the Raman image of the sample. Figure 3.15 shows the basic schematic of a Raman microscope.

Raman signals are intrinsically very weak, and conventional Raman imaging requires highly sensitive detectors. Specialized techniques such as coherent anti-Stokes Raman scattering (CARS), surface-enhanced Raman scattering (SERS), tip-enhanced Raman scattering (TERS), etc, have been developed to enhance the strength of the Raman signals and have been pivotal in the development and application of Raman imaging systems in biomedical sciences (Bhardwaj 2019). A detailed account of Raman spectroscopy is given in chapter 5.

3.5.3 Multispectral and hyperspectral imaging

The term spectral imaging is in general used to denote techniques that combine imaging with spectroscopy. This technique provides feature rich images, that can be used for precise characterization of the samples as well as enables automated identification or diagnosis. Based on the spectral resolution and spectral bands, spectral imaging techniques are sub-classified as multispectral, hyperspectral and ultraspectral imaging techniques. In spectral imaging, typically, the diffused reflection spectrum from each point on the sample is recorded. It is also possible to record the transmission spectral image, and even the fluorescence spectral image of the sample. It should be noted that proper referencing and radiometric calibration is necessary for accurate analysis of the multispectral and hyperspectral images recorded.

The data acquired from spectral imaging are usually represented in the form of multidimensional entities named as datacubes, which are the three-dimensional

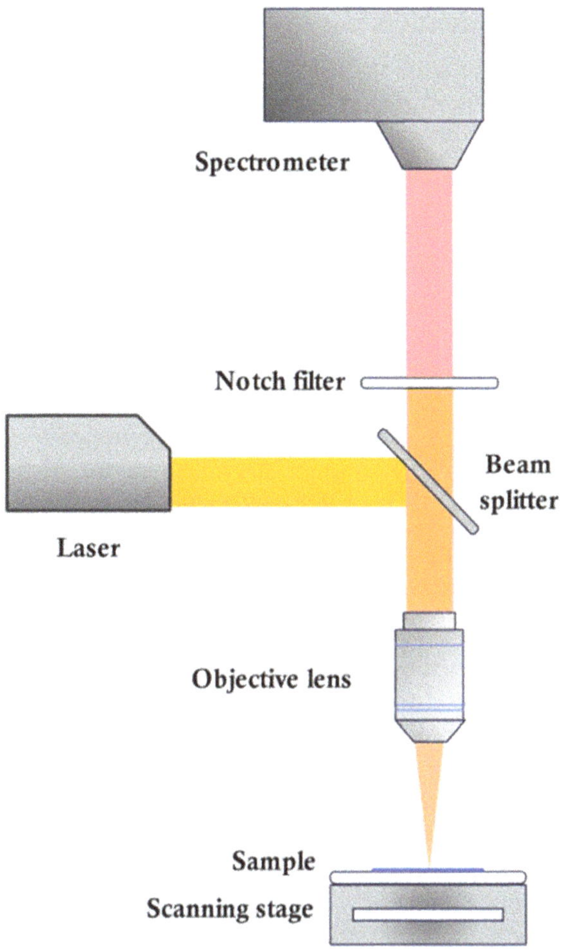

Figure 3.15. Schematic of a Raman imaging microscope.

representation of the acquired spatial-spatial-spectral data. Typically, the x- and y-axes represent the spatial data while the z-axis represents the spectral data. The spectral information about each point on the image shown in the xy-plane can be obtained from the z data corresponding to that point. The hyperspectral and ultraspectral datacubes contain huge amounts of data (big data). These datacubes are computationally intensive and it is challenging to analyse the data manually. To address this, dimension reduction and feature extraction tools are developed for specific analysis of these datacubes (Boldrini *et al* 2012). Multispectral and hyperspectral imaging are used widely in fields such as remote sensing, precision agriculture and biomedical imaging. Various multispectral and hyperspectral imaging modalities have been utilized for image-guided surgeries and disease diagnosis such as cancers, diabetic foot, retinal diseases and heart and circulatory pathology (Fei 2019). A detailed discussion on multispectral and hyperspectral imaging is presented in chapter 4.

3.6 Structured illumination microscopy (SIM)

Structured illumination microscopy (SIM) is an imaging technique that can provide spatial and axial resolutions greater than that of conventional wide-field imaging systems. As the name suggests, SIM uses structured patterns for illumination, rather than uniform illumination typically used in wide-field imaging. The basic idea behind SIM is that high spatial frequency information can be extracted from the sample using patterned illumination, and then the information can be computationally reconstructed to obtain higher resolution images. Structured illumination patterns used for SIM are typically generated either by the interference of coherent beams or by making use of spatial light modulators (SLMs). The structured illumination results in the generation of interference patterns through the moiré effect (Luxmoore and Shepherd 1983, Gustafsson 2000). When two periodic patterns having slightly different frequencies are superimposed one over the other or multiplied, the generated pattern is known as moiré fringe patterns. These patterns will generally have a frequency lower than either of the patterns. The shifted lower frequency of the moiré pattern can be resolved by the imaging system. Several images are recorded by rotating or translating the structured illumination and by deconvoluting the interference signals; extended resolution (in conventional SIM, up to two-fold improvement) images can be obtained.

The advantages of SIM include the resolution improvement, better image contrast, 3D-sectioning capability and the ability to record a wide field. The disadvantages include the requirement for recording multiple images for deconvolution, processing time required, and only two-fold resolution improvement. However, modern implementations of SIM such as saturated structured illumination microscopy (SSIM) and total internal reflection fluorescence SIM (TIRF-SIM) enable much higher resolutions (Chen *et al* 2023). The following section will give detailed discussion on this with specific aspects of periodic and aperiodic illumination beams.

3.6.1 Periodic illuminations

Structured light, also known as structured illumination, involves projecting a known pattern of light onto a scene. The primary goal of this technique is to detect and measure the deformation of the projected pattern on the surface of objects. This is commonly used in 3D reconstruction, where structured light systems project patterns (such as grids or horizontal bars) onto objects. The way these patterns distort when striking surfaces allows vision systems to calculate depth and surface details, as seen in structured light 3D scanners. Further, structured illumination microscopy (SIM) is a high-resolution imaging technique with significant potential in many industries, including healthcare.

3.6.1.1 Structured periodic illumination

In this method, a sinusoidal fringe pattern is projected onto the surface of the specimen through a microscope objective lens (see figure 3.16).

The surface topography affects the reflection or scattering of the pattern, which is then imaged onto a detector for 3D surface analysis. First developed by Engelhardt

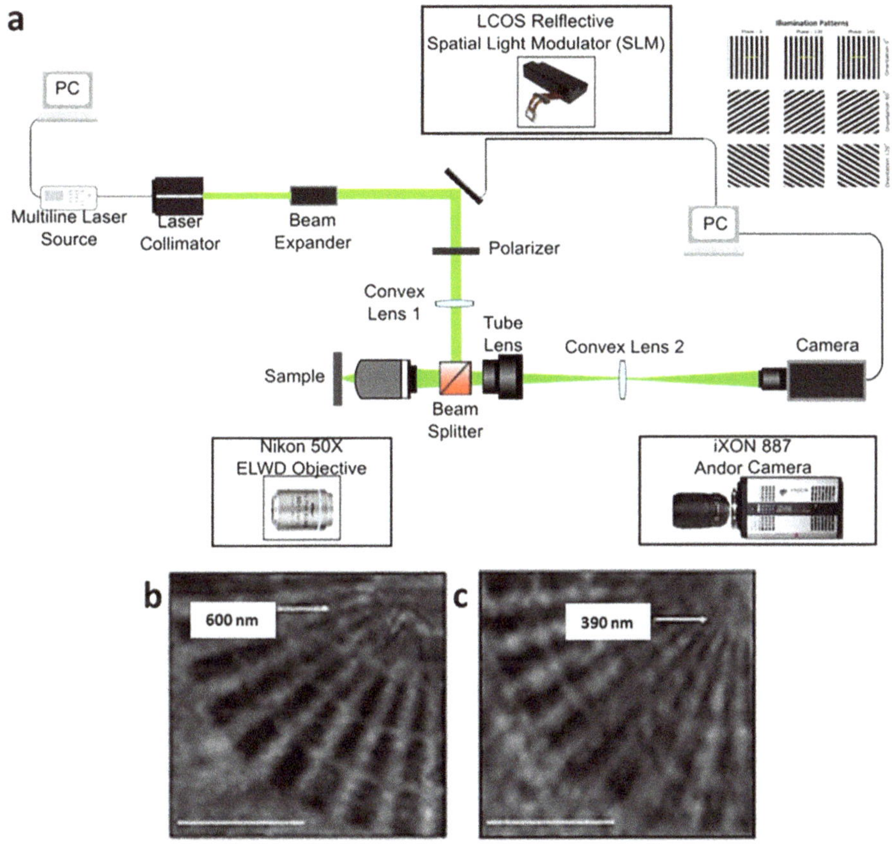

Figure 3.16. (a). Optical configuration of a SIM. Measurement of lateral resolution of conventional microscopy and SIM using a Siemen's star as the test sample is shown in (b), (c), respectively. The white scale bar indicates 7.5 μm. Adapted from Haridas *et al* (2020), copyright (2020) with permission from Elsevier.

and Häusler in 1988 for 3D surface topography, the technique was later enhanced by Neil and Wilson (Neil *et al* 1997). Despite its early development, SIM gained prominence when Gustafsson applied it to biological studies, demonstrating a two-fold improvement in lateral resolution (Komis *et al* 2015).

Several variations of SIM have emerged, but the most commonly used are super-resolution SIM (SR-SIM) and optical sectioning SIM (OS-SIM), which offer improved resolution and image clarity. These methods are now essential tools in fields requiring high-precision analysis.

3.6.1.2 Application in bio-imaging

In structured illumination microscopy (SIM), spatial resolution is enhanced by gathering information from frequency space beyond the observable region. This is achieved in reciprocal space, where the Fourier transform (FT) of a structured illumination image contains additional information from different areas of the frequency spectrum. By capturing several frames with shifted illumination phases, this data can be computationally separated and used to reconstruct a super-resolution image (Goodman and Cox 1969).

SIM microscopy has the potential to replace electron microscopy in some medical diagnoses, such as for kidney disorders, kidney cancer, and blood diseases. While the term 'structured illumination microscopy' gained popularity later, Guerra (1995) was the first to demonstrate the technique by utilizing patterned light to surpass the Abbe resolution limit, achieving an almost fivefold enhancement in resolution (Guerra 1995, Saxena *et al* 2015).

3.6.2 Aperiodic illumination (speckle illumination)

3.6.2.1 Speckles

Speckle refers to the random granular patterns observed when a highly coherent light source, such as a laser, illuminates an optically rough surface like paper, white paint, a display screen, or metal. Speckle patterns arise only when a monochromatic (coherent) light source interacts with surface height variations on the order of, or larger than, the wavelength of the illuminating light. When this type of surface is illuminated, the light's complex amplitude at any given point in space results from the superposition of multiple amplitude vectors with random phases. Due to the surface height variations, the resultant amplitude and phase change across different points, leading to the formation of a speckle pattern.

Speckle patterns are generally classified into two main types: objective and subjective speckles (Sirohi 2013). Objective speckle patterns form when a coherent light source illuminates a rough surface, and the scattered light is projected onto an observation plane without the use of an imaging lens. The average size of the objective speckles is defined by $1.22\ \lambda\ L/D$, where D is the object's diameter, L is the distance between the object and the observation plane, and λ is the wavelength of the illumination. In contrast, subjective speckle patterns arise when the illuminated surface is imaged onto an observation plane through a lens. The average speckle size for subjective speckles is given by the formula $1.22\ \lambda f/a$, where a is the aperture of the lens, f is the lens's focal length, and λ is the illumination wavelength. Figure 3.17(a) illustrates a subjective speckle pattern generated from a rough surface, while figures 3.17(b) and (c) depict the optical setups for generating objective and subjective speckle patterns, respectively.

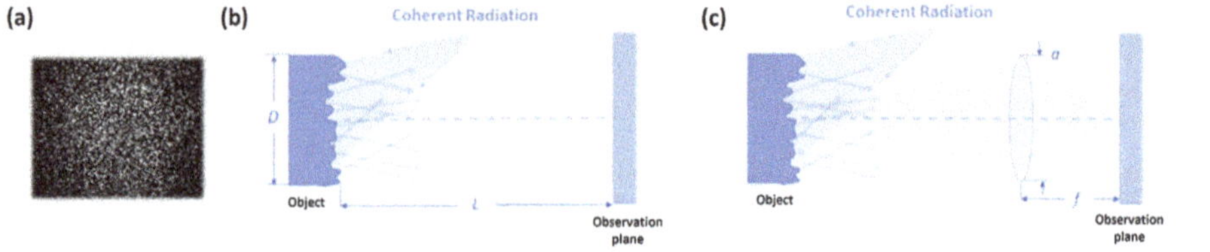

Figure 3.17. (a) A typical speckle pattern formed when a coherent light source illuminates a surface with optical roughness. The optical setups used to produce objective and subjective speckle patterns are illustrated in (b) and (c), respectively.

3.6.2.2 Structured illumination embedded speckle imaging

Aperiodic illumination pattern microscopy is an advanced imaging technique that improves spatial resolution by using non-repetitive, or aperiodic, illumination patterns to capture intricate details of biological specimens. Unlike traditional periodic illumination techniques, this method leverages the randomness of the pattern to gather more spatial frequency information from the sample. This allows the reconstruction of super-resolved images with enhanced detail beyond the diffraction limit. By illuminating the specimen with a series of aperiodic patterns and computationally reconstructing the resulting images, finer structural details, especially at the subcellular level, can be visualized.

This approach offers several advantages, particularly in biological microscopy, where it is essential to observe cellular processes at high resolution without damaging the specimen. Aperiodic illumination minimizes phototoxicity and photobleaching by distributing light more evenly across the sample compared to structured illumination microscopy (SIM). Additionally, it requires fewer acquisitions, which reduces the total imaging time. These features make aperiodic illumination pattern microscopy a powerful tool for live-cell imaging, enabling researchers to capture dynamic processes with minimal perturbation to the biological system (Ventalon and Mertz 2006, Hirakawa and Matsuki 2008, Mudry *et al* 2012).

Speckle patterns can be computer generated as shown in figures 3.18(a) and (b). These digital patterns, which appear as speckles embedded within the structured illumination patterns, have been shown to enhance spatial resolution (Haridas *et al* 2020).

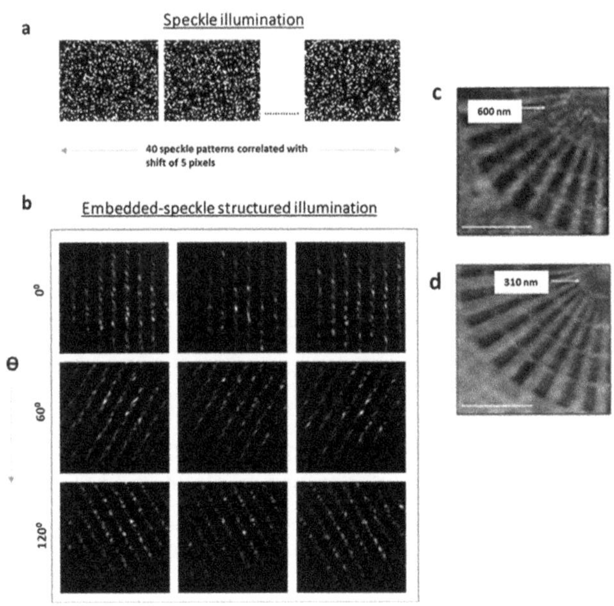

Figure 3.18. (a). Digital speckle patterns, (b) speckle patterns embedded into the structured illumination patterns, (c) represents image of a Siemen's star captured in wide field imaging mode, and (d) represents image captured using embedded speckle SIM. The white scale bar indicates 7.5 μm.

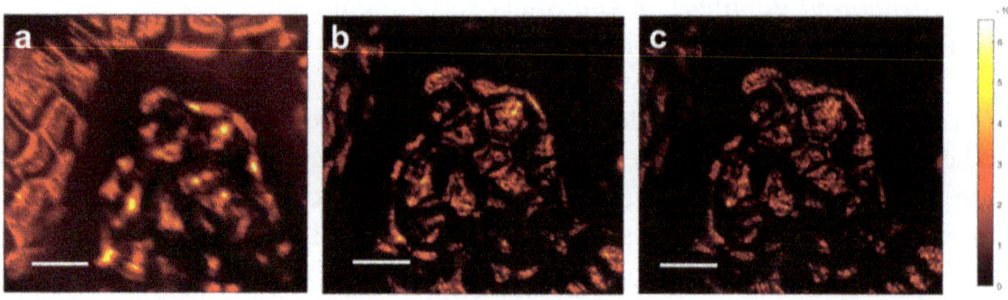

Figure 3.19. Image of necrosis in leaf sample captured in (a) conventional microscope mode. (b) Image acquired using structured illumination patterns. (c) Image captured by embedding speckles in the structured illumination patterns. The scale bar indicates 50 μm. Reproduced from Antony *et al* (2023) with permission from Elsevier.

In a study, a microscope that combines the advantages of structured illumination with embedded laser speckles (termed ES-SIM) was developed, introducing a technique named structured illumination embedded speckle microscopy (ES-SIM). This innovative approach offers sub-diffraction limit resolution imaging with a high signal-to-noise ratio (SNR). The speckle patterns act as pinholes, effectively reducing scattering noise, and applying the conventional SIM algorithm further enhances both imaging resolution and SNR. Using this optical configuration, an imaging resolution of approximately 310 ± 5 nm is reported—representing a 48% improvement over conventional microscopy—when viewed through a microscope objective (0.55 NA; 50×) with an 11 mm working distance, using a Siemen's star as the test sample (figures 3.18 (c) and (d)) (Haridas *et al* 2020). Additionally, the mean square error (MSE) improved by around 45%. The study also revealed that while the resolution improvement is independent of the speckle size, the enhancement in SNR is strongly influenced by it. The developed high-resolution, high-SNR microscope is anticipated to drive significant advancements in engineering applications, potentially reshaping current methodologies in the field. A biological application using such a system for detecting necrotic regions has been reported recently (Antony *et al* 2023). Microscopic images of a *Piper Sarmentosum* leaf sample with necrotic lesions, captured using three different imaging techniques, are presented in figure 3.19. It is evident that SIM and ES-SIM provide high-contrast images compared to a conventional microscope, enabling accurate necrotic region detection.

3.7 Problems

1. Explain the working principle of a phase contrast microscope.
2. How does a differential interference contrast microscope enhance image contrast?
3. Name three imaging systems that can subdue the limitations of wide-field imaging systems.
4. How does a chromatic confocal imaging system differ from normal confocal imaging system.
5. Describe the major difference between a SD-OCT and a SS-OCT system.

6. What are speckles? Describe how speckles can be embedded in a structured illumination microscope.
7. What is periodic and aperiodic illumination? Explain each with an example.
8. How can embedding speckles in a SIM improve the resolution? Give a detailed analysis.

References

Antony M M, Haridas A, Suchand Sandeep C S and Vadakke Matham M 2023 An optodigital system for visualizing the leaf epidermal surface using embedded speckle SIM: a 3D non-destructive approach *Comput. Electron. Agric.* **211** 107962

Aumann S, Donner S, Fischer J and Müller F 2019 Optical Coherence Tomography (OCT): principle and technical realization ed J F Bille *High Resolution Imaging in Microscopy and Ophthalmology* (Cham: Springer International Publishing) pp 59–85

Beard P 2011 Biomedical photoacoustic imaging *Interface Focus* **1** 602–31

Berezin M Y and Achilefu S 2010 Fluorescence lifetime measurements and biological imaging *Chem. Rev.* **110** 2641–84

Bhardwaj V 2019 Surface-enhanced Raman spectroscopy *Nanotechnology Science and Technology* (New York: Nova Science Publishers)

Boldrini B, Kessler W, Rebner K and Kessler R W 2012 Hyperspectral imaging: a review of best practice, performance and pitfalls for in-line and on-line applications *J. Near Infrared Spectrosc.* **20** 483–508

Born M, Wolf E, Bhatia A B, Clemmow P C, Gabor D, Stokes A R, Taylor A M, Wayman P A and Wilcock W L 1999 *Principles of Optics: Electromagnetic Theory of Propagation, Interference and Diffraction of Light* 7th edn (Cambridge: Cambridge University Press)

Cabaleiro P, De Moura J, Novo J, Charlón P and Ortega M 2019 Automatic identification and representation of the cornea–contact lens relationship using AS-OCT images *Sensors* **19** 5087

Chen X, Zhong S, Hou Y, Cao R, Wang W, Li D, Dai Q, Kim D and Xi P 2023 Superresolution structured illumination microscopy reconstruction algorithms: a review *Light: Sci. Appl.* **12** 172

Chorvat D and Chorvatova A 2009 Multi-wavelength fluorescence lifetime spectroscopy: a new approach to the study of endogenous fluorescence in living cells and tissues *Laser Phys. Lett.* **6** 175–93

Coons A H, Creech H J, Jones R N and Berliner E 1942 The demonstration of pneumococcal antigen in tissues by the use of fluorescent antibody *J. Immunol.* **45** 159–70

Corydon T J *et al* 2016 Alterations of the cytoskeleton in human cells in space proved by life-cell imaging *Sci. Rep.* **6** 20043

Drexler W and Fujimoto J G (ed) 2015 *Optical Coherence Tomography: Technology and Applications* (Cham: Springer International Publishing)

Drexler W, Liu M, Kumar A, Kamali T, Unterhuber A and Leitgeb R A 2014 Optical coherence tomography today: speed, contrast, and multimodality *J. Biomed. Opt.* **19** 071412

Dubois A and Boccara C 2006 L'OCT plein champ *Med. Sci.* **22** 859–64

Farooq A, Sabah S, Dhou S, Alsawaftah N and Husseini G 2022 Exogenous contrast agents in photoacoustic imaging: an in vivo review for tumor imaging *Nanomaterials* **12** 393

Fei B 2019 Hyperspectral imaging in medical applications *Data Handling in Science and Technology* (Amsterdam: Elsevier) pp 523–65

Girach A and Sergott R C (ed) 2016 *Optical Coherence Tomography* (Cham: Springer International Publishing)

Grieve K, Dubois A, Simonutti M, Paques M, Sahel J, Le Gargasson J-F and Boccara C 2005 In vivo anterior segment imaging in the rat eye with high speed white light full-field optical coherence tomography *Opt. Express* **13** 6286

Goodman J W and Cox M E 1969 Introduction to Fourier optics *Phys. Today* **22** 97–101

Guerra J M 1995 *Super-resolution through illumination by diffraction-born evanescent waves Appl. Phys. Lett.* **66** 3555–7

Gustafsson M G L 2000 Surpassing the lateral resolution limit by a factor of two using structured illumination microscopy: short communication *J. Microsc.* **198** 82–7

Haridas A, Perinchery S M, Shinde A, Buchnev O and Murukeshan V M 2020 Long working distance high resolution reflective sample imaging via structured embedded speckle illumination *Opt. Lasers Eng.* **134** 106296

Hirakawa Y and Matsuki Y 2008 Living plant observation by a laser speckle microscopy *Rev. Laser Eng.* **36** 1355–7

Hrebesh M S, Dabu R and Sato M 2009 In vivo imaging of dynamic biological specimen by real-time single-shot full-field optical coherence tomography *Opt. Commun.* **282** 674–83

Huang D *et al* 1991 Optical coherence tomography *Science* **254** 1178–81

James J and Tanke H J 1991 *Biomedical Light Microscopy* (Dordrecht: Springer)

Jensen E C 2012 Types of imaging, part 2: an overview of fluorescence microscopy *Anat. Rec.* **295** 1621–7

Jerome W G and Price R L (ed) 2018 *Basic Confocal Microscopy* (Cham: Springer International Publishing)

Komis G, Samajova O, Ovecka M and Samaj J 2015 Super-resolution microscopy in plant cell imaging *Trends Plant Sci.* **20** 834–43

Leroux C-E, Palmier J, Boccara A C, Cappello G and Monnier S 2015 Elastography of multicellular aggregates unpublished to osmo-mechanical stress *New J. Phys.* **17** 073035

Luxmoore A R and Shepherd A T 1983 Applications of the Moiré effect ed A R Luxmoore *Optical Transducers and Techniques in Engineering Measurement* (Dordrecht: Springer) pp 61–108

Mondal P P and Diaspro A 2014 *Fundamentals of Fluorescence Microscopy: Exploring Life with Light* (Dordrecht: Springer)

Mudry E, Belkebir K, Girard J, Savatier J, Le Moal E, Nicoletti C, Allain M and Sentenac A 2012 Structured illumination microscopy using unknown speckle patterns *Nat. Photonics* **6** 312–5

Neil M A, Juskaitis R and Wilson T 1997 Method of obtaining optical sectioning by using structured light in a conventional microscope *Opt. Lett.* **22** 1905–7

Rasband W S 1997–2018 ImageJ, U. S. National Institutes of Health, Bethesda, Maryland, USA, https://imagej.net/ij/

Raveendran S, Lim H-T, Maekawa T, Vadakke Matham M and Sakthi Kumar D 2018 Gold nanocages entering into the realm of high-contrast photoacoustic ocular imaging *Nanoscale* **10** 13959–68

Ruzin S E 2024 Differential interference contrast microscopy, modulation contrast microscopy *Techniques in Light Microscopy* (Oxford: Oxford University Press) pp 89–104

Saxena M, Eluru G and Gorthi S S 2015 Structured illumination microscopy *Adv. Opt. Photonics* **7** 241–75

Shen C and Zhang Y 2023 Staining technology and bright-field microscope use *Food Microbiology Laboratory for the Food Science Student* (Cham: Springer International Publishing) pp 9–17

St. Croix C M, Shand S H and Watkins S C 2005 Confocal microscopy: comparisons, applications, and problems *BioTechniques* **39** 1

Sirohi R S 2013 *Speckle Metrology* (Taylor and Francis)

Suhling K *et al* 2015 Fluorescence lifetime imaging (FLIM): basic concepts and some recent developments *Med. Photonics* **27** 3–40

Swedlow J R and Platani M 2002 Live cell imaging using wide-field microscopy and deconvolution *Cell Struct. Funct.* **27** 335–41

Turrell, G and Corset, J 1996 *Raman Microscopy* (Amsterdam: Elsevier)

Ventalon C and Mertz J 2006 Dynamic speckle illumination microscopy with translated versus randomized speckle patterns *Opt. Express* **14** 7198–209

Wang L V (ed) 2017 *Photoacoustic Imaging and Spectroscopy* 1st edn (Boca Raton, FL: CRC Press)

Yu Q, Wang C, Zhang Y, Hu S, Liu T, Cheng F, Wang Y, Lin T and Xi L 2022 Error analysis and correction of thickness measurement for transparent specimens based on chromatic confocal microscopy with inclined illumination *Photonics* **9** 155

Zoubir A (ed) 2012 *Raman Imaging: Techniques and Applications* Springer Series in Optical Sciences (Berlin: Springer)

IOP Publishing

Diagnostic Biomedical Optics
Fundamentals and applications
Murukeshan Vadakke Matham, C S Suchand Sandeep, Maria Merin Antony, Manojit Pramanik and Santhosh Chidangil

Chapter 4

Multispectral and hyperspectral imaging

Maria Merin Antony and Murukeshan Vadakke Matham

This chapter provides an in-depth exploration of multispectral and hyperspectral imaging (HSI), focusing on the principles, instrumentation, and applications of these technologies in various diagnostic imaging domains. The chapter is divided into several sections, each addressing key aspects of multispectral and hyperspectral imaging systems. Instrumentation of the HSI system is discussed in detail, covering the essential steps for setting up the equipment. This includes the spectral calibration of the CCD camera, which ensures accurate wavelength measurements, and the spatial calibration, which focuses on precise spatial alignment of the imaging system. The chapter also addresses the processes involved in data acquisition and processing, which are crucial for extracting meaningful information from HSI data. A section on resolution explains the importance of achieving high spatial and spectral clarity in HSI, which is essential for accurate analysis.

4.1 Introduction

Spectral imaging is an advanced technology that merges conventional imaging with spectroscopy to capture both spatial and spectral information about an object (Park *et al* 2000). Its origins date back nearly four decades to the launch of Landsat 1 in 1972. This method captures images across hundreds of continuous wavelength or spectral bands, allowing a unique radiance spectrum to be obtained for each pixel. Initially, in the early 1990s, spectral imaging was mainly applied in areas such as terrestrial military operations, remote sensing, and astrophysics. However, what began in remote sensing and metrology has since expanded into numerous scientific and technological fields. This growth is due to the technology's ability to provide a wide range of highly detailed and specific spectral data. The benefits of spectral imaging, including multispectral (MS) and hyperspectral (HS) techniques, compared to traditional imaging methods like monochrome, RGB, and standard spectroscopy, are summarized in table 4.1 and illustrated in figure 4.1.

doi:10.1088/978-0-7503-2364-2ch4

Table 4.1. Comparison of monochrome, RGB, spectroscopy, multispectral, and hyperspectral features. Adapted with permission from Li *et al* (2013) under the license CC BY 4.0.

Feature	Monochrome	RGB	Spectroscopy	MS	HS
Spatial information	Yes	Yes	No	Yes	Yes
Number of spectral bands	1	3	100 s	3–10	100 s
Spectral information	No	No	Yes	Limited	Yes
Multiconstituent information	No	Limited	Yes	Limited	Yes
Sensitivity to minor components	No	No	No	Limited	Yes

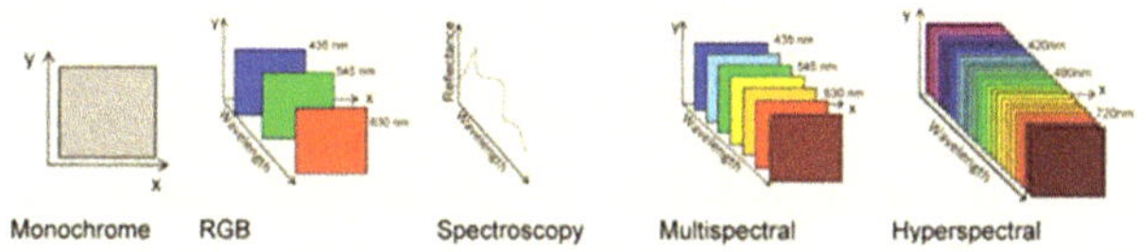

Figure 4.1. Diagrammatic representation of monochrome, RGB, spectroscopy, multispectral and hyperspectral features. Adapted with permission from Li *et al* (2013) under the license CC BY 4.0.

As listed in table 4.1, the hyperspectral and multispectral imaging can provide more sensitive spectral and spatial information about the object imaged compared to conventional monochrome, RGB, and spectroscopic techniques.

4.2 Multispectral and hyperspectral imaging

Multispectral (MS), hyperspectral (HS), and ultraspectral (US) imaging are commonly used terms in spectral imaging, distinguished by the increasing number of spectral bands and the precision of data capture. However, there is no universally accepted guidelines that exist to distinctly differentiate between these terms. Table 4.1 presents two classification criteria, both of which use the number of spectral bands as a key defining parameter. Fresse *et al* (2010) focus on precision, while Puschell (Park *et al* 2000) emphasize resolution in their respective definitions. By using the number of wavelength bands, a shared criterion between these approaches, the latter definition simplifies the classification of a system as either HS or US (tables 4.2 and 4.3).

With the detection of more spectral bands, an expanded spectral range, and improvements in precision and resolution, hyperspectral and ultraspectral imaging provide more detailed spectral signatures, which can be used for highly accurate identification purposes. The fine spectral resolution of ultraspectral imaging allows for capturing molecular absorption or emission bands, making it a powerful tool for precise spectroscopic analysis (Meigs *et al* 1998).

Table 4.2. Comparison of spectral imaging techniques according to Fresse's definition. Data obtained from (Fresse *et al* 2010, Lim H-T and Murukeshan V M 2015).

Imaging Category	Typical Band Count	Relative Spectral Resolution ($\Delta\lambda/\lambda$)
MS	5–10 bands	~0.1 (coarse spectral detail)
HS	100–200 bands	~0.01 (fine spectral detail)
US	1,000–10,000 bands	~0.001 (ultra-fine spectral resolution)

Table 4.3. Comparison of spectral imaging techniques according to Puschell's definition. Data obtained from (Park *et al* 2000, Lim H-T and Murukeshan V M 2015).

Imaging category	Approximate Band Range	Typical Spectral Resolution (nm)
MS	5–20 spectral bands	~20–100 nm
HS	30–300 bands	$\leqslant$10 nm
US	$\geqslant$300 bands	$\leqslant$1 nm

4.3 Hyperspectral imaging

Hyperspectral imaging (HSI) has been a significant tool in airborne and spaceborne remote sensing since 1989 (Keith Hege *et al* 2004), marking a shift from traditional film-based methods to advanced electronic recording systems (Antony *et al* 2024). This technological shift made it possible to capture narrow, contiguous spectral bands over an extensive spectral range, allowing for detailed spectral information to be collected for each spatial pixel. The spectral data acquired can be matched with known material spectra, enabling the classification and measurement of different substances using established data libraries or even the identification of unknown materials. Initially, HSI was primarily utilized in remote sensing (Goetz *et al* 1985, Thenkabail *et al* 2012), but its applications have since expanded to various domains (Antony *et al* 2019, 2022), such as quality control of agricultural products (Gao *et al* 2006, Mahesh *et al* 2015, Merin Antony *et al* 2020, 2023, Merin Antony *et al* 2024), biomedical imaging (Lu and Fei 2014, Lim and Murukeshan 2016a, 2016b), and forensic analysis (Lim and Murukeshan 2017, Melit Devassy and George 2020).

HSI systems are specifically designed to deliver high spectral resolution across chosen spectral ranges, making them highly suitable for imaging multiple fluorescence tags concurrently. Even when these tags have overlapping emission spectra, HSI can effectively distinguish them through spectral analysis. This capability overcomes a significant challenge of traditional spectroscopic imaging, which relies on minimal spectral overlap and requires separate band-pass filters for each tag. By using HSI, there is greater flexibility in the number and combinations of tags that can be employed in imaging.

4.3.1 Datacube

HSI generates a datacube, which is a three-dimensional (3D) representation that captures both spatial and spectral information. A cutaway view of the datacube, as shown in figure 4.2, reveals its internal structure. From this datacube, multiple two-dimensional (2D) spatial images, each corresponding to a specific spectral band, can be extracted. Each voxel within the datacube contains intensity data for a specific spectral band at a single spatial point in the 2D sample (Lu and Fei 2014, Antony *et al* 2024). A spectrum for any given spatial point is generated by extracting data along the spectral axis, providing detailed spectral information from adjacent narrow bands. This rich spectral data can be analyzed for classification and quantification by comparing it with reference libraries using various algorithms.

There are several methods available for performing HSI, which can be categorized into three main categories: spatial scanning (whiskbroom and pushbroom), spectral scanning, and snapshot imaging. These methods vary in how they acquire data to construct the datacube (figure 4.2). Each technique offers specific advantages and limitations, making the choice of method dependent on the particular requirements and applications of the imaging task.

4.3.2 Types of HSI techniques

Various HSI techniques are discussed, including point-scan (whiskbroom), line-scan (pushbroom), spectral scanning, and snapshot imaging, each with distinct advantages and applications (Antony *et al* 2024). The different hyperspectral imaging configurations are shown in figure 4.3.

Spatial-scanning HSI systems are commonly employed in both laboratory and field environments. These systems typically utilize a dispersive element, such as a prism–grating–prism configuration in a spectrograph, to break down incoming light into its constituent spectral bands, which are then detected by a sensor array. Spatial-scanning imagers can be categorized into two main categories: whiskbroom (point-scan) and pushbroom (line-scan) imagers. Some of the spectral imaging systems are equipped with video cameras to enable direct video imaging.

A whiskbroom HSI imager conducts point scanning, capturing the spectrum at one spatial point per scan, resulting in one-dimensional (1D) spectral data. By scanning across multiple points in a 2D area, a datacube is created. This spatial scanning can occur either by moving the sample with a 2D stage or by directing point illumination through a micro-electro-mechanical system (MEMS) scanner. While this method is effective, it can be time-consuming for larger samples since

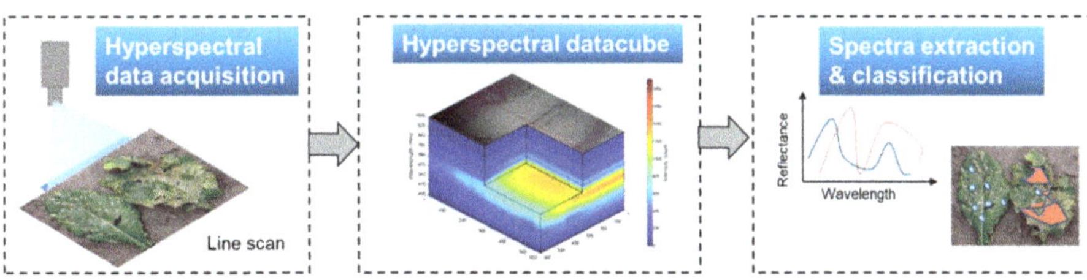

Figure 4.2. Hyperspectral datacube.

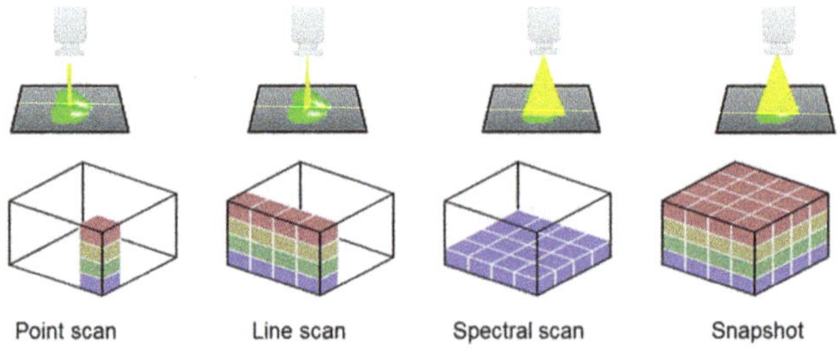

Figure 4.3. Different hyperspectral imaging configurations.

each point requires a separate scan. In contrast, the line-scanning pushbroom imager (figure 4.3) captures spectral data from an entire line of the sample during each scan. Light from this line passes through a narrow slit, gets dispersed into various wavelengths, and is recorded by a 2D sensor array. This process is repeated after moving the sample relative to the imager, building the datacube. In comparison to the point-scanning method, the line-scanning pushbroom imager is faster and more efficient, as it collects more information with each scan.

Spectral-scanning HSI imagers are also used in both table-top and field settings (Gat 2000). These systems employ electronically tunable filters, such as acousto-optical tunable filters (AOTF) or liquid crystal tunable filters (LCTF), to regulate the transmitted wavelengths (Guan *et al* 2011). Each scan yields a 2D spatial image at a specific spectral band. By adjusting the tunable filter to allow different spectral bands, multiple images can be obtained to construct the 3D datacube. The AOTF can switch between wavelengths in under 1 ms, making it considerably faster than the LCTF, which is more suitable for high-frame-rate video applications (Leitner *et al* 2010). Unlike spatial-scanning imagers, spectral-scanning systems allow users to modify the number of captured wavelength bands, thereby reducing acquisition time when fewer bands are needed. Additionally, no movement is required between the sample and detector for each scan.

When comparing acquisition times between line-scanning and spectral-scanning imagers, the primary factor affecting line-scanning systems is the number of rows and the switching time between them, which depends on distance and motion speed. In contrast, acquisition time for spectral-scanning imagers is influenced by the number of spectral bands and the switching times. When the sample remains stationary, both imager types can accurately capture the spectrum at each point. However, unexpected movement during scanning affects the data differently in each system. In spatial-scanning imagers, sample motion during row switching may misalign the spatial data while maintaining spectral accuracy. Conversely, spectral-scanning imagers will record inaccurate spectral data for all points if the sample moves unexpectedly. Therefore, spectral-scanning systems require stricter stability during acquisition, especially when capturing multiple wavelength bands, although image registration algorithms can help correct any distortions (Gat 2000).

Snapshot HSI systems are often utilized in both table-top and field applications (Lim and Murukeshan 2016a, Ren and Allington-Smith 2002). These systems can acquire the entire 3D datacube in a single scan using configurations such as image

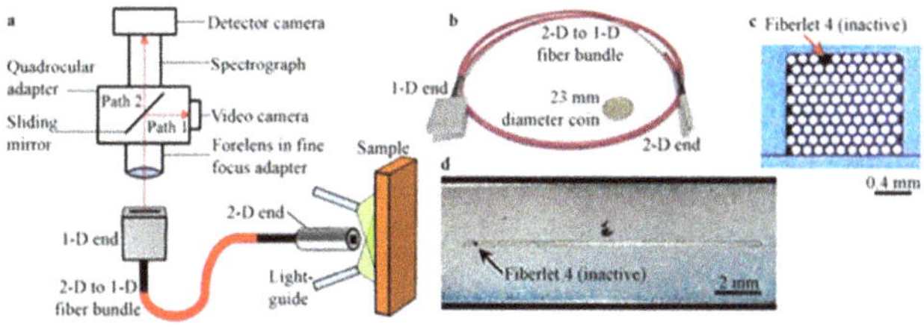

Figure 4.4. (a) Schematic diagram of a probe based snapshot HSI system. (b) Photograph of the 2D-1D fiber probe. (c) 2D end face of the fiber probe. (d) 1D end face of the fiber probe. Reproduced with permission from Lim and Murukeshan (2016a) under the license CC BY 4.0.

mapping spectroscopy, integral field spectroscopy, computed tomographic imaging spectroscopy, and compressive sensing (Ren and Allington-Smith 2002, Hagen *et al* 2012, Gao *et al* 2012, Lim and Murukeshan 2016a, Vanderriest 2016). Unlike spatial- and spectral-scanning imagers, snapshot systems do not rely on sequential scanning. The main advantage of snapshot imagers is their speed, making them suitable for real-time applications, subject to exposure time and detector readout rates. This capability minimizes motion artifacts and pixel misregistration. However, because the 2D detector has a limited pixel count, snapshot imagers can only capture a finite amount of data in one scan, often necessitating a trade-off between the number of spatial points and the number of wavelengths recorded.

One common configuration for snapshot HSI is integral field spectroscopy, which employs a reformatter, such as a fiber bundle, box, or rod (Ren and Allington-Smith 2002). The fiberlets on one end of the reformatter are arranged in a 2D array, while the opposite end is organized in a 1D row (Hagen *et al* 2012). Light from the sample is gathered by the 2D fiberlet array and transmitted to the 1D end, where the fiberlets function as a slit for a spectrograph-based HSI system. This configuration enables the 2D sensor array to capture 3D spatial–spectral data (figure 4.4). Data processing is necessary to rearrange the spectra according to the positions of the fiberlets in the 2D array, allowing for accurate visualization of the data. Integral field spectroscopy has seen widespread use in both table-top and field applications across various fields, including astronomy and ocular imaging (Leitner *et al* 2010, Gao *et al* 2012, Vanderriest 2016).

4.4 Instrumentation of the HSI system

This section covers the technical components and instrumentation involved in building a hyperspectral imaging system. It discusses light sources, optics, detectors (e.g., CCD cameras), and the importance of filters for capturing narrow spectral bands. The importance of spectral calibration is discussed in this section, focusing on how to ensure the camera accurately captures spectral data.

4.4.1 Spectral calibration of the CCD camera

The point light source with known spectral lines is used as a calibration source and is dispersed as it passes through the spectrograph along the DC_Y direction. A second

Table 4.4. Wavelength calibration using ArHg source.

Real wavelength (nm)	Pixel position along DC_Y
404.656	110
435.833	152
544.074	296
578.000	336
694.543	486
704.722	500

order polynomial model was used to relate each DC_Y to its calibration wavelength as shown in equation (4.1), where *a*, *b*, and *c* are constants. With these constants, each DC_Y was assigned a wavelength (λ).

$$\lambda = a \bullet DC_Y^2 + b \bullet DC_Y + c \tag{4.1}$$

where *a*, *b* and *c* are constants. An example of a calibration for an ArHg source is shown in table 4.4.

Using table 4.4, the values of the constants *a*, *b*, *c* can be calculated as $a = 0.000\,073\,4536$, $b = 0.725\,977$, $c = 331.871$. For $\lambda = 400$ nm, the value of DC_Y is given by 93 and for $\lambda = 1000$ nm, the value of DC_Y is given by 848. The number of bands is given by the difference between the DCY for 1000 and 400 nm = 756 bands.

4.4.2 Spatial calibration

Spatial calibration is a crucial aspect of hyperspectral imaging systems, ensuring that spectral data are accurately mapped to spatial positions, particularly in real-time applications. This process is essential for both spatial scanning and snapshot hyperspectral imagers.

4.4.2.1 Spatial scanning hyperspectral imager

Spatial scanning HSI systems, including point-scan and line-scan systems, rely on mechanical stages for precise movement. For instance, the pushbroom hyperspectral (HS) imager sequentially scans the region of interest (ROI) from top to bottom. The movement of the *y*-axis stage between consecutive scans is determined by a parameter called 'Step.' For instance, when the Step value is set to 5, the *y*-axis stage advances by a distance corresponding to five rows of the sensor array in the camera. A larger Step value results in faster acquisition times but can reduce the *y*-axis spatial resolution. Therefore, the Step parameter must be carefully adjusted to achieve an optimal balance between data acquisition speed and spatial resolution.

4.4.2.2 Snapshot hyperspectral imager

This section explores an example of a snapshot HSI system that utilizes a 2D-to-1D converter. The system described is a 4D snapshot hyperspectral imager with a wavelength range of 400–1000 nm, capable of detecting 756 wavelength bands, which is significantly higher than many comparable systems. The fourth dimension, time, is introduced by capturing sequential three-dimensional datacubes. The system is composed of two main components: a hyperspectral imager and a 2D-to-1D fiber bundle (figures 4.4(a)–(d)).

The 2D end of the fiber bundle captures a two-dimensional image and compresses it into a single dimension at its 1D end. Light from the 1D end is then fed into a spectrograph, where it is dispersed and detected by a 2D sensor array. This fiber bundle simplifies the three-dimensional (spatial–spatial–spectral) data into two dimensions (spatial–spectral), allowing the 2D sensor array to capture all the necessary information in a single scan.

The data is recorded sequentially in real-time, introducing the fourth dimension, time, into the dataset. After acquisition, a custom MATLAB code processes each acquired spectrum, mapping it back to its original spatial position on the 2D end of the fiber bundle as illustrated in figures 4.5(a)–(c).

The spatial calibration of the 1D end of the fiber bundle was performed following its alignment and positioning. The fiberlets were sequentially numbered from 1 to 100, starting from the leftmost corner. The detector camera, equipped with a sensor array containing 1002 rows (y-axis for spectral data) and 1004 columns (x-axis for spatial data), captured 2D spectral-spatial data from the fiberlets located at the 1D end of the fiber bundle. In figure 4.5(a), each colored vertical line represents light emerging from the core of a fiberlet, which was then spectrally dispersed along the y-axis of the sensor array. The y-axis corresponds to calibrated spectral bands, while the position of the colored lines along the x-axis indicates the pixel columns imaging of each fiberlet.

During data processing, the spectral information from each fiberlet was extracted from the corresponding pixel columns. Figure 4.5(a) also displays the camera's

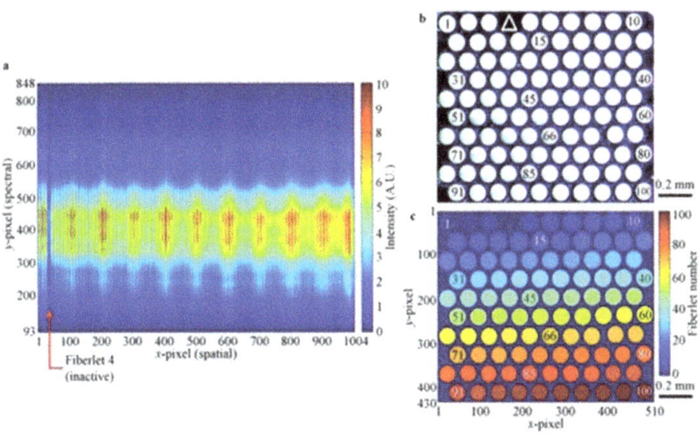

Figure 4.5. (a) Reference image along the 1D end captured by the detector camera, (b) photograph of the fiberlets on the 2D end-face, and (c) digital mask of the 2D end-face. The numbers on some of the fiberlets in (b) and (c) represent their respective numbering. The white triangle in (b) marks the position of Fiberlet 4, which was inactive and therefore appeared dark. Reproduced with permission from Lim and Murukeshan (2016a) under the license CC BY 4.0.

average reference frame after dark compensation. A digital representation of the 2D end-face was generated, as illustrated in figures 4.5(b) and (c), displaying the positioning and numbering of the fiberlets.

4.4.2.3 Spectral scanning hyperspectral imager

In spectral scanning systems, spatial calibration is generally unnecessary because the system primarily focuses on scanning the spectral domain rather than the spatial domain. Unlike spatial scanning hyperspectral imagers, which require precise movement across the region of interest to gather spatial data, spectral scanning systems collect information across a range of wavelengths for a fixed spatial position. The scanning process occurs along the spectral dimension, capturing detailed spectral information without the need to reposition or scan across different spatial areas. This inherent design eliminates the need for spatial calibration, as the spatial positions remain fixed while the system varies the wavelength to gather spectral data. Therefore, the accuracy of these systems is primarily dependent on spectral calibration rather than spatial calibration.

4.5 Data processing

The data processing system is responsible for capturing and analyzing the spectral data obtained from the hyperspectral camera. This process generally involves three key stages: datacube generation, data preprocessing, and data classification. The specific steps in the initial data processing stage depend on the type of hyperspectral imaging configuration being used, such as pushbroom, whiskbroom, spectral scan, or snapshot. The system includes software capable of handling the large data volume generated by HSI, as well as the tools needed for analysis, such as: spectral angle mapper, principal component analysis (PCA), machine learning techniques.

- Spectral angle mapper (SAM): a widely used algorithm for classification that compares the angle between the spectral signature of a pixel and reference spectra in a hyperspectral datacube.
- PCA: reduces the dimensionality of the hyperspectral data while preserving essential information, making it easier to interpret the spectral variance across the sample.
- Machine learning techniques: for advanced data processing and classification, HSI systems often incorporate machine learning models, such as Random Forest, SVM, or deep learning algorithms, to automate the analysis and enhance predictive accuracy.

The final output of an HSI system is a hyperspectral datacube, which contains spatial information (x, y) and spectral information (λ) for every pixel in the image. The datacube is analyzed to extract spectral signatures that correspond to the chemical or physical properties of the material. Visual representation: the hyperspectral images can be displayed as RGB composites or grayscale images based on specific wavelengths or band ratios to highlight key features of the sample. Each pixel's spectral signature is compared to a reference library or subjected to algorithms (e.g., SAM) to classify materials, detect anomalies, or monitor changes over time.

4.6 Resolution

The resolution of any HSI system is defined in terms of spectral resolution and spatial resolution.

Spectral Resolution: The hyperspectral camera divides incoming light into hundreds of narrow spectral bands, usually with a high spectral resolution (2–10 nm). This enables the detection of subtle differences in the reflectance, absorbance, or transmittance properties of materials.

Spatial Resolution: The camera simultaneously captures spatial information, recording the spatial distribution of spectral signatures within the scene. Spatial resolution can vary depending on the optics and sensor configuration, allowing detailed mapping at different scales. In spatial scanning systems, the spatial resolution additionally depends on the step size of the scanning stage.

4.7 Diagnostic imaging using HSI

Over the decades, HSI has emerged as a powerful tool in the field of diagnostic imaging due to its ability to capture detailed wavelength information across a wide spectral range, beyond the visible spectrum. Unlike traditional imaging methods, which capture only RGB color information, HSI acquires hundreds of spectral bands for each pixel in an image, providing a wealth of data that can be used to distinguish between different types of tissues and detect subtle changes that may indicate disease. In medical diagnostics, HSI offers significant advantages. It enables non-invasive imaging with high sensitivity to biochemical and structural variations in tissues. This makes it particularly useful for identifying early-stage diseases, such as cancer, where the metabolic and molecular changes precede visible symptoms. HSI can analyze tissue composition by detecting specific absorption and reflection patterns related to key biomolecules like hemoglobin, water, and lipids. One of the key applications of HSI in diagnostics is in cancer detection, where it is used to differentiate between healthy and cancerous tissues based on their unique spectral signatures. This can be applied in various procedures, including tumor margin delineation during surgery, ensuring that cancerous cells are accurately removed while preserving healthy tissue (see figure 4.6). A table listing the hyperspectral imaging, its characteristics and applications is summarized in table 4.5.

HSI is also being explored in ophthalmology for diagnosing retinal diseases, dermatology for skin lesion analysis, and gastroenterology for detecting abnormal tissue in endoscopic procedures. HSI's ability to provide functional information, such as blood oxygenation levels, tissue perfusion, and water content, further enhances its diagnostic utility. Diagnostic applications of HSI can be broadly divided into: Disease detection and surgical guidance which are specifically classified based on the diseases and surgical procedures (see figure 4.4).

Surgical guidance applications, such as endoscopy, necessitate a snapshot configuration due to the requirement for real-time data acquisition. In this regard,

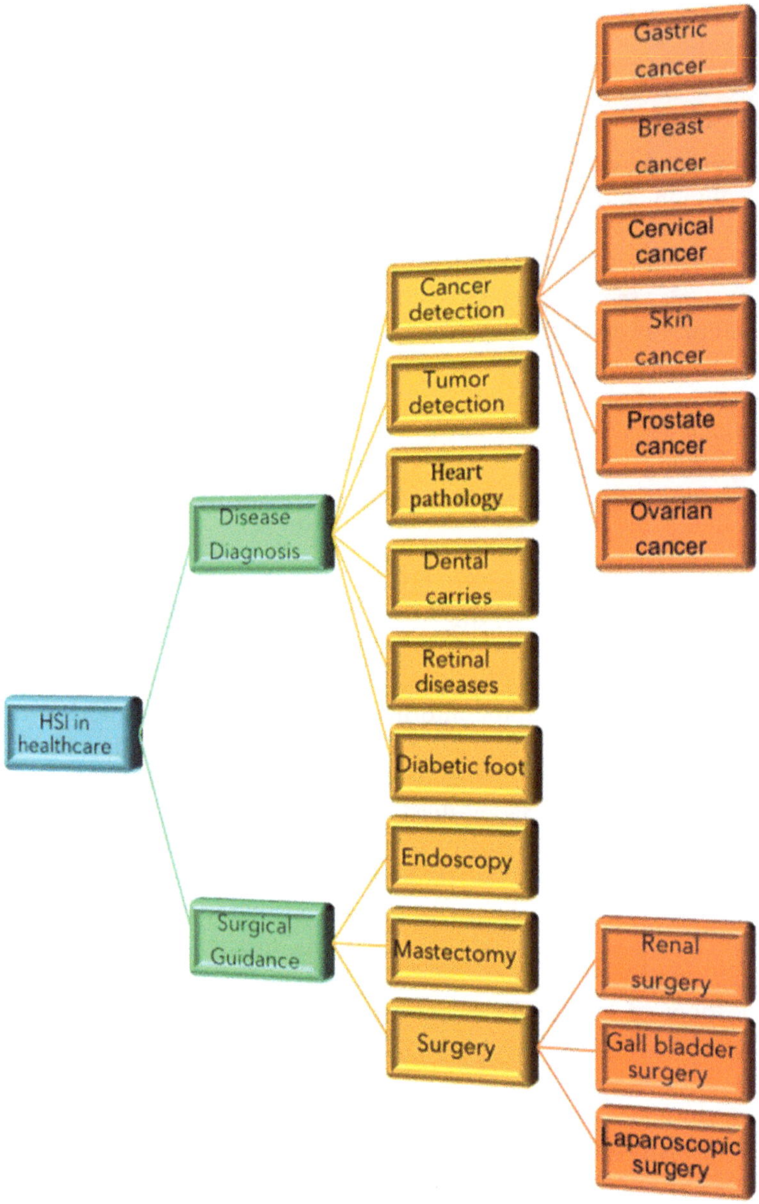

Figure 4.6. Hyperspectral imaging in diagnostic applications.

the performance of the endoscopic probe described in section 4.4.2, developed for real-time operation, is presented in figure 4.7.

Using this video-endoscope, the tumor region in human tissues can be accurately and sensitively identified based on the extracted reflectance spectrum. By integrating HSI with AI/machine learning algorithms, automatic classification and detection of

Table 4.5. Summary of HSI systems and its applications. R denotes Reflectance, T denotes Transmission, F denotes Fluorescence. Adapted from Lu and Fei (2014) under license CC BY 4.0.

Spectral range (nm)	Detector used	Dispersive element	Mode of data acquisition	Measurement mode (R, F,T)	Applications	References
400–1100	Si CCD	Filter wheel	Wavelength/ spectral scan	R	Burn wounds	Afromowitz *et al* (1988)
200–700	CCD	Filter wheel	Wavelength/ spectral scan	F and R	Cervical neoplasia	Ferris *et al* (2001)
330–480	CCD	Filter wheel	Wavelength/ spectral scan	F and R	Cervical cancer	Benavides *et al* (2003)
530–680	CCD	Prism	Pushbroom/line scan	T	Cutaneous wound	Shah *et al* (2003)
500–600	CCD	LCTF	Wavelength/ spectral scan	R	Diabetic foot	Greenman *et al* (2005)
400–720	CCD	LCTF	Wavelength/ spectral scan	F	Tumor hypoxia and microvasculature	Sorg *et al* (2005)
440–640	CCD; ICCD	AOTF	Wavelength/ spectral scan	F and R	Skin cancer	Kong *et al* (2006)
500–600	CCD	LCTF	Wavelength/ spectral scan	R	Haemorrhagic shock	Cancio *et al* (2006)
365–800	CCD	Prism	Pushbroom/line scan	T	Melanoma	Dicker *et al* (2006)
400–1000	Si CCD	Grating	Pushbroom/line scan	R	Skin bruises	Randeberg *et al* (2006)
900–1700	InGaAs					Randeberg and Hernandez-Palacios (2012)
950–2500	HgCdTe					Randeberg and Hernandez-Palacios (2012)
450–700	FPA	CGH	Snapshot	R	Ophthalmology	Johnson *et al* (2007)
450–700	CCD	LCTF	Wavelength/ spectral scan	R	Breast cancer	Panasyuk *et al* (2007)

650–1100	FPA	LCTF	Wavelength/ spectral scan	R	Laparoscopic surgery	Zuzak *et al* (2007)
400–1000; 900–1700	CCD; InGaAs	PGP	Pushbroom/line scan	R	Intestinal ischemia	Akbari *et al* (2010)
1000–2500	HgCdTe	PGP	Pushbroom/line scan	R	Gastric cancer	Akbari *et al* (2011)
450–650	CCD	Prism	Snapshot	R	Endoscope	Kester *et al* (2011)
410–1000	Si CCD	Grating	Pushbroom/line scan	R and F	Atherosclerosis	Larsen *et al* (2011)
400–720	CCD	LCTF	Wavelength/ spectral scan	R	Diabetic foot	Yudovsky *et al* (2011)
450–950	CCD	LCTF	Wavelength/ spectral scan	R	Prostate cancer	Akbari *et al* (2012)
390–680	CCD	Grating	Pushbroom/line scan	R	Laryngeal disorders	Martin *et al* (2012)
400–640	CCD	Filter wheel	Wavelength/ spectral scan	F and R	Ovarian cancer	Renkoski *et al* (2012)
900–1700	InGaAs	AOTF	Wavelength/ spectral scan	R	Dental caries	Rechmann *et al* (2012)
550–950	CCD	AOTF	Wavelength/ spectral scan	T	Leucocyte pathology	Guan(2012)
400–1000	CCD	2D–1D fiber bundle	Snapshot	R, F	Endoscopy applications	Lim and Murukeshan (2016a)

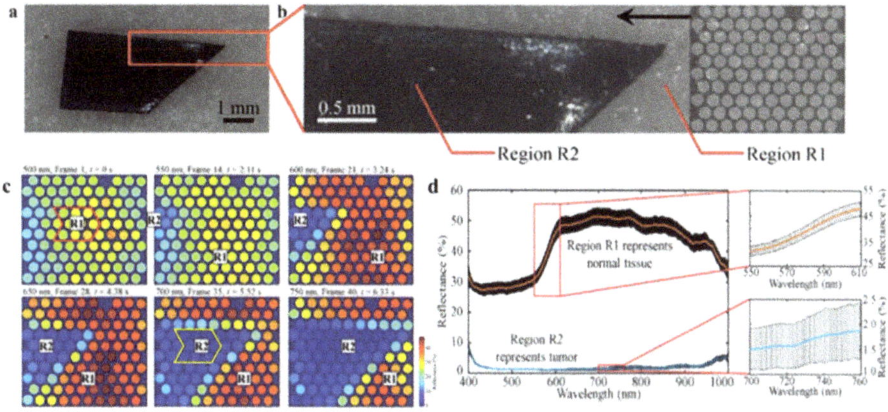

Figure 4.7. (a) Phantom tissue sample containing a simulated tumor. (b) Images showing the sample and the 2D end of the fiber bundle, with region R1 representing the phantom tissue and region R2 indicating the simulated tumor area. (c) HS reflectance maps derived from the datacubes of the sample at various spectral bands over time. (d) Reflectance spectra obtained from the two sample regions (R1 and R2). Adapted from Lim and Murukeshan (2016a) with permission under the CC BY 4.0 license.

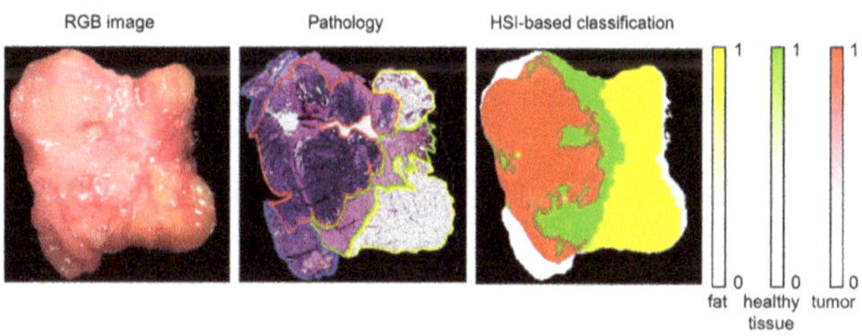

Figure 4.8. HSI-based classification performed via the trained artificial model. Adapted with permission from Baltussen *et al* (2019) under license CC BY 4.0.

pathological conditions can be achieved, making it a valuable tool for clinical decision-making as shown in figure 4.8 (Baltussen *et al* 2019).

SLIMBRAIN represents a cutting-edge AR system designed specifically for the real-time acquisition, processing, and classification of hyperspectral data to assist in the identification of brain tumour tissue during surgical procedures as shown in figure 4.9. The primary goal of this system is to provide neurosurgeons with immediate and precise visual feedback, allowing them to distinguish between healthy and cancerous brain tissue during tumour resections.

In contrast to RGB image, the HS image captured using the SLIMBRAIN can provide the real-time classification of cancerous tissue enabling accurate removal of the tissue (see figure 4.9(b)). The system operates by capturing hyperspectral images at an impressive rate of 14 frames per second (fps), which is critical for maintaining real-time performance in a dynamic surgical environment. This rapid image acquisition and processing capability ensures that the detection and delineation of cancerous tissue occur simultaneously as the surgeon carries out the operation, minimizing delays and enhancing surgical precision.

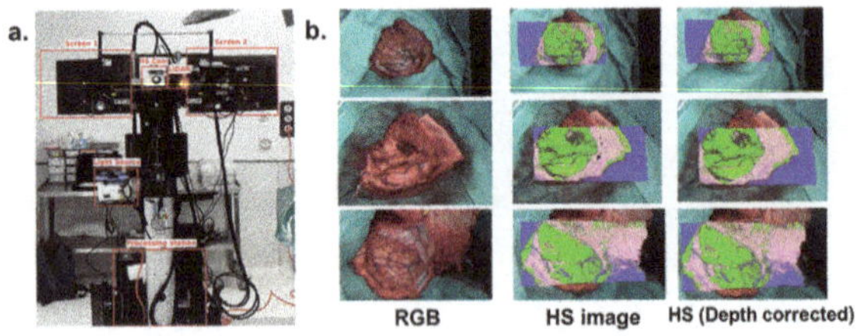

Figure 4.9. (a) Schematic diagram of the SLIMBRAIN system. (b) RGB images and corresponding HS-classified images. Adapted with permission from Sancho *et al* (2023) under license CC BY 4.0.

The evaluation of various HSI modalities and systems remains limited due to the absence of standardized assessment methods. Optical systems used in spectral imaging exhibit variations in spatial, spectral, and temporal resolution, which complicates the establishment of universal HSI standards. To address this issue, the introduction of standard tissue-mimicking phantoms has recently been proposed. These phantoms offer a quantitative means of evaluating optical systems, facilitating the standardization of HSI technologies.

In addition to these challenges, the implementation of quantitative 3D hyperspectral imaging in clinical environments presents significant difficulties. The ability to acquire 3D imaging through optical techniques is essential for tasks such as defining lesion boundaries and diagnosing metastasis. However, conventional HSI systems typically capture broadband light signals via spectroscopic approaches, which results in a loss of critical 3D depth information. Although cutting tissue into thin slices, similar to histological tissue preparation, can facilitate 3D hyperspectral imaging, this method is impractical for rapid image acquisition in clinical settings.

Although HSI holds great promise, several challenges impede its widespread use in clinical settings. Key issues include the need for standardized imaging protocols, efficient real-time data processing, and the development of more compact and user-friendly devices. However, as HSI technology continues to advance, it is expected to significantly contribute to precision medicine by offering clinicians a more detailed understanding of tissue health, ultimately leading to earlier and more accurate diagnoses (Lim and Murukeshan 2016b). The limitations of traditional HSI, can be overcome using innovative techniques such as spectral optical coherence tomography and multispectral photoacoustic. These techniques enable the acquisition of 3D spectral data with all the necessary attributes. Despite their potential, these approaches still require further improvements in optical system design to enhance their feasibility and promote wider clinical adoption (Antony *et al* 2024).

4.8 Problems

1. What is spectral imaging? Explain the advantages it offers for diagnostic imaging applications.
2. What are the different types of spectral imaging?
3. With the help of a schematic diagram, explain briefly the instrumentation details of a hyperspectral imaging system.

4. Explain the concept of a hyperspectral datacube with suitable examples.
5. What are the different types of hyperspectral imaging?
6. Explain briefly how hyperspectral camera is spectrally calibrated.
7. What are the data processing techniques used in hyperspectral imaging?
8. With illustrative examples, explain how hyperspectral imaging is used in biomedical diagnostic applications.

References

Afromowitz M A, Callis J B, Heimbach D M, DeSoto L A and Norton M K 1988 Multispectral imaging of burn wounds: a new clinical instrument for evaluating burn depth *IEEE Trans. Biomed. Eng.* **35** 842–50

Akbari H, Halig L V, Schuster D M, Osunkoya A, Master V, Nieh P T, Chen G Z and Fei B 2012 Hyperspectral imaging and quantitative analysis for prostate cancer detection *J. Biomed. Opt.* **17** 076005

Akbari H, Kosugi Y, Kojima K and Tanaka N 2010 Detection and analysis of the intestinal ischemia using visible and invisible hyperspectral imaging *IEEE Trans. Biomed. Eng.* **57** 2011–7

Akbari H, Uto K, Kosugi Y, Kojima K and Tanaka N 2011 Cancer detection using infrared hyperspectral imaging *Cancer Sci.* **102** 852–7

Antony M M, Sandeep S, Lim C S, H-T and Vadakke Matham M 2022 Hyperspectral vision based probe for in situ corrosion monitoring in saline environments *IEEE Trans. Instrum. Meas.* **71** 1–7

Antony M M, Suchand Sandeep C S and Matham M V 2019 Monitoring system for corrosion in metal structures using a probe based hyperspectral imager *Proc. SPIE. 11205, 7th Int. Conf. on Optical and Photonic Engineering 2019* (Bellingham, WA: SPIE)

Antony M M, Suchand Sandeep C S, Lim H-T and Vadakke Matham M 2023 High-resolution ultra-spectral imager for advanced imaging in agriculture and biomedical applications *J. Biomed. Photon. Eng.* **9** 8928

Antony M M, Suchand Sandeep C S and Vadakke Matham M 2024 Hyperspectral vision beyond 3D: a review *Opt. Lasers Eng.* **178** 108238

Baltussen E J M *et al* 2019 Hyperspectral imaging for tissue classification, a way toward smart laparoscopic colorectal surgery *J. Biomed. Opt.* **24** 1–9

Benavides J, Chang S, Park S, Richards-Kortum R, Mackinnon N, Macaulay C, Milbourne A, Malpica A and Follen M 2003 Multispectral digital colposcopy for in vivo detection of cervical cancer *Opt. Express* **11** 1223–36

Cancio L C, Batchinsky A I, Mansfield J R, Panasyuk S, Hetz K, Martini D, Jordan B S, Tracey B and Freeman J E 2006 Hyperspectral imaging: a new approach to the diagnosis of hemorrhagic shock *J. Trauma* **60** 1087–95

Dicker D T, Lerner J, Van Belle P, Barth S F, Guerry D T, Herlyn M, Elder D E and El-Deiry W S 2006 Differentiation of normal skin and melanoma using high resolution hyperspectral imaging *Cancer Biol. Ther.* **5** 1033–8

Ferris D G, Lawhead R A, Dickman E D, Holtzapple N, Miller J A, Grogan S, Bambot S, Agrawal A and Faupel M L 2001 Multimodal hyperspectral imaging for the noninvasive diagnosis of cervical neoplasia *J. Low Genit. Tract. Dis.* **5** 65–72

Fresse V, Houzet D and Gravier C 2010 GPU architecture evaluation for multispectral and hyperspectral image analysis *2010 Conf. on Design and Architectures for Signal and Image Processing (DASIP)* (Piscataway, NJ: IEEE) 121–7

Gat N 2000 Imaging spectroscopy using tunable filters: a review *Wavelet Applications VII, 50–64, Proc. SPIE* (Bellingham, WA: SPIE) 4056

Gao L, Smith R T and Tkaczyk T S 2012 Snapshot hyperspectral retinal camera with the Image Mapping Spectrometer (IMS) *Biomed. Opt. Express* **3** 48–54

Goetz A F, Vane G, Solomon J E and Rock B N 1985 Imaging spectrometry for earth remote sensing *Science* **228** 1147–53

Greenman R L, Panasyuk S, Wang X, Lyons T E, Dinh T, Longoria L, Giurini J M, Freeman J, Khaodhiar L and Veves A 2005 Early changes in the skin microcirculation and muscle metabolism of the diabetic foot *Lancet* **366** 1711–7

Guan Y 2012 Pathological leucocyte segmentation algorithm based on hyperspectral imaging technique *Opt. Eng.* **51** 053202

Guan Y, Li Q, Liu H, Xu L and Zhu Z 2011 New-styled system based on hyperspectral imaging *Symp. on Photonics and Optoelectronics (SOPO)* (Piscataway, NJ: IEEE) 1–3

Hagen N, Kester R T, Gao L and Tkaczyk T S 2012 Snapshot advantage: a review of the light collection improvement for parallel high-dimensional measurement systems *Opt. Eng.* **51** 111702

Johnson W R, Wilson D W, Fink W, Humayun M and Bearman G 2007 Snapshot hyperspectral imaging in ophthalmology *J. Biomed. Opt.* **12** 014036

Keith Hege E, O'Connell D, Johnson W, Basty S and Dereniak E L 2004 Hyperspectral imaging for astronomy and space surviellance *Proc. Vol 5159, Imaging Spectrometry IX* https://doi.org/10.1117/12.506426

Kester R T, Bedard N, Gao L and Tkaczyk T S 2011 Real-time snapshot hyperspectral imaging endoscope *J. Biomed. Opt.* **16** 056005

Kong S G, Martin M E and Vo-Dinh T 2006 Hyperspectral fluorescence imaging for mouse skin tumor detection *ETRI J.* **28** 770–6

Larsen E L, Randeberg L L, Olstad E, Haugen O A, Aksnes A and Svaasand L O 2011 Hyperspectral imaging of atherosclerotic plaques in vitro *J. Biomed. Opt.* **16** 026011

Leitner R, Arnold T and De Biasio M 2010 High-sensitivity hyperspectral imager for biomedical video diagnostic applications *Proc. SPIE 7674, Smart Biomedical and Physiological Sensor Technologies VII* **7674** 76740E

Li Q, He X, Wang Y, Liu H, Xu D and Guo F 2013 Review of spectral imaging technology in biomedical engineering: achievements and challenges *J. Biomed. Opt.* **18** 100901

Lim H-T and Murukeshan V M 2015 Pushbroom hyperspectral imaging system with selectable region of interest for medical imaging *J. Biomed. Opt.* **20** 046010

Lim H-T and Murukeshan V M 2017 Hyperspectral imaging of polymer banknotes for building and analysis of spectral library *Opt. Lasers Eng.* **98** 168–75

Lim H T and Murukeshan V M 2016a A four-dimensional snapshot hyperspectral video-endoscope for bio-imaging applications *Sci. Rep.* **6** 24044

Lim H T and Murukeshan V M 2016b Spatial-scanning hyperspectral imaging probe for bio-imaging applications *Rev. Sci. Instrum.* **87** 033707

Lu G and Fei B 2014 Medical hyperspectral imaging: a review *J. Biomed. Opt.* **19** 10901

Mahesh S, Jayas D S, Paliwal J and White N D G 2015 Hyperspectral imaging to classify and monitor quality of agricultural materials *J. Stored Prod. Res.* **61** 17–26

Martin R, Thies B and Gerstner A O 2012 Hyperspectral hybrid method classification for detecting altered mucosa of the human larynx *Int. J. Health Geogr.* **11** 21

Meigs A D, Otten L J and Cherezova T Y 1998 Ultraspectral imaging: a new contribution to global virtual presence *IEEE Aerospace Conf. Proc. (Cat. No.98TH8339)* **2** 5–12
Melit Devassy B and George S 2020 Dimensionality reduction and visualisation of hyperspectral ink data using t-SNE *Forensic Sci. Int.* **311** 110194
Merin Antony M, Suchand Sandeep C S and Matham M V 2020 Probe-based hyperspectral imager for crop monitoring *Proc. SPIE 11525, SPIE Future Sensing Technologies* (Bellingham, WA: SPIE) 1152512
Merin Antony M, Suchand Sandeep C S, Bijeesh M M and Matham M V 2024 A fast analysis approach for crop health monitoring in hydroponic farms using hyperspectral imaging *Proc. SPIE 12879, Photonic Technologies in Plant and Agricultural Science* (Bellingham, WA: SPIE) 128790G
Michael L W, Susan L U, Pablo Z-T, Alicia P-O and Vern C V 2006 Hyperspectral mapping of crop and soils for precision agriculture *Proc. SPIE 6298, Remote Sensing and Modeling of Ecosystems for Sustainability III* (Bellingham, WA: SPIE) 62980B
Panasyuk S V, Yang S, Faller D V, Ngo D, Lew R A, Freeman J E and Rogers A E 2007 Medical hyperspectral imaging to facilitate residual tumor identification during surgery *Cancer Biol. Ther.* **6** 439–46
Park S K, Puschell J J and Rahman Z-u 2000 Hyperspectral imagers for current and future missions *Visual Inform. Process.* **IX** 121–32
Randeberg L L and Hernandez-Palacios J 2012 Hyperspectral imaging of bruises in the SWIR spectral region *Proc. SPIE 8207, Photonic Therapeutics and Diagnostics VIII* 82070N
Randeberg L L *et al* 2006 Hyperspectral imaging of bruised skin *Proc. SPIE 6078, Photonic Therapeutics and Diagnostics II* 60780O
Ren D and Allington-Smith J 2002 On the application of integral field unit design theory for imaging spectroscopy *Publ. Astron. Soc. Pac.* **114** 866–78
Renkoski T E, Hatch K D and Utzinger U 2012 Wide-field spectral imaging of human ovary autofluorescence and oncologic diagnosis via previously collected probe data *J. Biomed. Opt.* **17** 036003
Sancho J, Villa M, Chavarrías M, Juarez E, Lagares A and Sanz C 2023 SLIMBRAIN: augmented reality real-time acquisition and processing system for hyperspectral classification mapping with depth information for in vivo surgical procedures *J. Syst. Archit.* **140** 102893
Shah S A, Bachrach N, Spear S J, Letbetter D S, Stone R A, Dhir R, Prichard J W, Brown H G and LaFramboise W A 2003 Cutaneous wound analysis using hyperspectral imaging *Biotechniques* **34** 408–13
Sorg B S, Moeller B J, Donovan O, Cao Y and Dewhirst M W 2005 Hyperspectral imaging of hemoglobin saturation in tumor microvasculature and tumor hypoxia development *J. Biomed. Opt.* **10** 44004
Thenkabail P S and Huete J G L A 2012 *Hyperspectral Remote Sensing of Vegetation* 1st edn (New York: CRC Press Taylor and Francis Group)
Vanderriest C 2016 Integral field spectroscopy with optical fibres *Int. Astronom. Union Coll.* **149** 209–18
Yudovsky D, Nouvong A, Schomacker K and Pilon L 2011 Assessing diabetic foot ulcer development risk with hyperspectral tissue oximetry *J. Biomed. Opt.* **16** 026009
Zuzak K J, Naik S C, Alexandrakis G, Hawkins D, Behbehani K and Livingston E H 2007 Characterization of a near-infrared laparoscopic hyperspectral imaging system for minimally invasive surgery *Anal. Chem.* **79** 4709–15

IOP Publishing

Diagnostic Biomedical Optics

Fundamentals and applications

Murukeshan Vadakke Matham, C S Suchand Sandeep, Maria Merin Antony, Manojit Pramanik and Santhosh Chidangil

Chapter 5

Raman spectroscopy techniques for medical applications: analysis of human blood components

N Mithun, K P Sreejith and Santhosh Chidangil

The chapter discusses the most popular Raman spectroscopy techniques, such as micro-Raman spectroscopy, surface-enhanced Raman spectroscopy and Raman Tweezers spectroscopy, and some of their medical applications. These techniques offer valuable insights into important biomolecular species in human clinical samples and their structure and functions under different health conditions. The clinical samples selected for the study consist of blood components, cells and tissues. Raman-based techniques to be discussed here are non-destructive and require a minimal quantity of samples, and sample preparation is relatively easy, making it advantageous for various analytical applications.

5.1 Introduction

5.1.1 Fundamentals of Raman spectroscopy

Raman scattering is a phenomenon that occurs when light interacts with matter, resulting in a change in the energy (and thus the wavelength) of the scattered light. Sir CV Raman was awarded the Nobel Prize in Physics in 1930 for this ground-breaking discovery (Raman 1928). Raman scattering is an inelastic scattering phenomenon where incident light interacts with molecular vibrations or rotations, causing the scattered light to shift in frequency (Miles *et al* 2001, Jayasooriya and Jenkins 2002). This frequency shift provides information about the vibrational modes of the molecule. The interaction between light and molecules results in two types of scattering: significant elastic scattering (Rayleigh scattering) and faint inelastic scattering (Raman scattering).

doi:10.1088/978-0-7503-2364-2ch5

Molecular polarization plays a crucial role in the Raman scattering process. It refers to the process by which a molecule develops an electric dipole moment when exposed to an external electric field. Polarization arises due to the rearrangement of electron density within the molecule, causing positive and negative charges to shift relative to one another. This shift results in a temporary dipole moment even in molecules that are otherwise non-polar. The degree of polarization depends on the molecular structure and the strength of the external field. Polarization is important in understanding many molecular interactions and responses, especially in electromagnetic fields, affecting how light interacts with matter. In Raman spectroscopy, the ability of a molecule to scatter light in response to incident radiation is influenced by the polarization of both the light and the molecule's polarizability. The polarizability of a molecule can change as it vibrates, and these changes in polarizability give rise to Raman scattering.

The molecular polarizability α describes how easily the electron cloud of a molecule can be distorted by an external electric field E, leading to the induction of a dipole moment P.

The equation for molecular polarizability is

$$P = \alpha E \tag{5.1}$$

where p is the induced dipole moment (in units of Cm), α is the molecular polarizability (in units of $C{\cdot}m^2\ V^{-1}$) and E is the external electric field (in units of $V\ m^{-1}$).

5.1.1.1 Polarizability of small molecules

For symmetric diatomic molecules (like H_2, N_2, or O_2), the polarizability depends on how easily their electron cloud is distorted by an external electric field. Diatomic molecules are typically non-polar, but their polarizability can change during molecular vibrations, which is essential for Raman scattering. In a symmetric diatomic molecule, the only vibrational mode is the stretching mode, where the bond length between the two atoms changes periodically. The critical factor determining whether this vibration is Raman active is whether the polarizability changes during this vibration. In contrast to IR spectroscopy, where the dipole moment must change, in Raman spectroscopy it is the polarizability that must change during vibration for a mode to appear in the spectrum.

In the Raman spectrum of a symmetric diatomic molecule, one can observe a single Raman active vibrational mode corresponding to the stretching of the bond between the two atoms. This is due to changes in polarizability when light interacts with this molecule. Typical Raman shifts for diatomic molecules O_2, N_2 and H_2 are $\sim$1556, $\sim$2331, and $\sim$4160 cm^{-1}, respectively.

The water molecule (H_2O) is an excellent example for understanding its vibrational modes. Water, a non-linear molecule, has three fundamental vibrational modes, which can be active in both infrared (IR) and Raman spectroscopy. Water has three main vibrational modes due to its molecular structure.

(1) Symmetric stretching (ν_1): in this mode, both O–H bonds stretch or compress symmetrically. The oxygen atom remains stationary while both hydrogen atoms move in and out. The frequency of this vibration comes out around 3650 cm^{-1}.

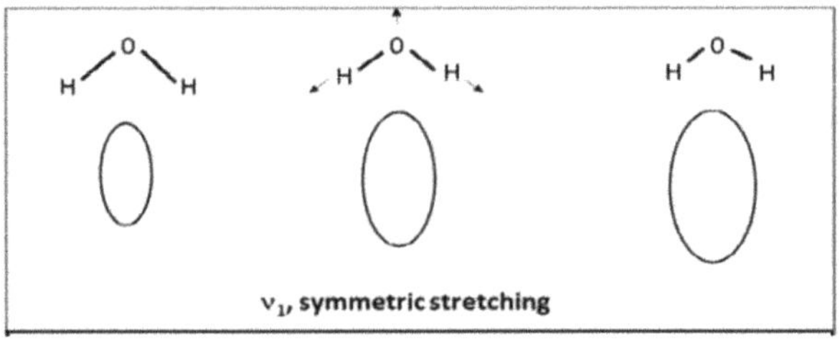

Figure 5.1. Symmetric stretching and polarizability ellipsoids of water molecule.

(2) Asymmetric stretching (ν_3): one O–H bond stretches while the other compresses. The hydrogen atoms move in opposite directions, and the corresponding frequency is around 3756 cm^{-1}. (3) Bending or scissoring (ν_2): its frequency is 1595 cm^{-1}. In this vibrational mode, the H atoms move closer together or further apart while the oxygen atom remains stationary. In the Raman spectrum of water, peaks corresponding to these vibrational modes appear due to the inelastic scattering of light as the vibrational states change. The symmetric stretching and bending modes are prominent in Raman spectra because they involve significant changes in molecular polarizability. Polarizability ellipsoids for symmetric vibrational modes are shown in figure 5.1.

5.1.1.2 Classical theory of Raman effect

In the classical description of Raman scattering, incident light is treated as an oscillating electromagnetic wave, and its interaction with molecules is described in terms of the induced dipole moment caused by the oscillating electric field of the light. This induced dipole is the source of scattered radiation. The strength of the induced dipole is proportional to the molecular polarizability, which depends on how easily the electron cloud in the molecule is distorted by the electric field. This is given by:

$$P(t) = \alpha E(t) \tag{5.2}$$

where $P(t)$ is the induced dipole moment, α is the polarizability of the molecule and $E(t)$ is the time-dependent electric field of the light wave.

Light can be described as an oscillating electric field:

$$E = E_0 \sin(2\pi\nu t) \tag{5.3}$$

where E_0 is the amplitude of the electric field and ν is the frequency of the incident light.

A polarizable molecule in an electric field will have an induced dipole that oscillates with the frequency of the light, given by the relation:

$$P(t) = a[E_0 \sin(2\pi\nu t)] \tag{5.4}$$

In addition to interacting with the oscillating electric field, the molecule undergoes vibrational motion. If the vibrational frequency is denoted by ν_{vib}, the molecular polarizability changes periodically with the vibrational motion of the molecule. This can be expressed as:

$$\alpha = \alpha_0 + \beta \sin(2\pi\nu_{vib}t) \tag{5.5}$$

where α_0 is static polarizability and β change in polarizability.

Combining equations (5.4) and (5.5):

$$\begin{aligned} P(t) &= [\alpha_0 + \beta \sin(2\pi\nu_{vib}t)] \times [E_0 \sin(2\pi\nu t)] \\ &= [\alpha_0 E_0 \sin(2\pi\nu t)] + [\beta E_0 \sin(2\pi\nu_{vib}t)\sin(2\pi\nu t)] \\ &= [\alpha_0 E_0 \sin(2\pi\nu t)] + \tfrac{1}{2}\beta E_0[\cos(2\pi(\nu - \nu_{vib})t) - \cos(2\pi(\nu + \nu_{vib})t)] \end{aligned} \tag{5.6}$$

Equation (5.6) contains three key terms. The term $\alpha_0 E_0 \sin(2\pi\nu t)$ represents Rayleigh scattering, which is elastic scattering. In this case, the scattered light has the same frequency as the incident light. The term $\cos(2\pi(\nu-\nu_{vib})t)$ represents Stokes–Raman scattering, where the scattered light has a lower frequency than the incident light frequency. The term $\cos(2\pi(\nu\nu_{vib})t)$ represents anti-Stokes–Raman scattering, where the scattered light has a higher frequency than the incident light. In the classical theory, a vibrational mode is Raman active if the molecular polarizability changes during the vibration.

The classical theory of the Raman effect provides a useful and intuitive framework for understanding light–molecule interactions. The classical theory assumes continuous vibrational energy changes in the molecule. However, molecular vibrational energy levels are quantized, meaning that molecules can only absorb or emit light at specific energy levels corresponding to allowed vibrational transitions. The classical theory fails to explain selection rules that determine which vibrational modes are Raman active or inactive. Classical theory cannot distinguish between Stokes and anti-Stokes scattering. It does not provide a way to predict the relative intensities of Raman lines. The classical theory primarily focuses on vibrational transitions and does not adequately explain rotational Raman scattering, which occurs due to changes in rotational energy levels. In this theory, polarizability changes during molecular vibrations are considered in a simplistic, linear manner, which can limit the accuracy of predicting the Raman activity of more complex molecules. It cannot explain resonance Raman effect. The classical theory does not account for the influence of temperature on the relative intensities of Stokes and anti-Stokes lines, nor does it explain why anti-Stokes lines are typically weaker.

Raman spectroscopy is often preferred over infrared (IR) spectroscopy for biological studies due to several advantages related to sample handling, spectral information, and compatibility with biological environments.

5.1.1.3 Quantum theory of Raman effect

When the light photons interact with molecules, they will initially be in the lower vibrational energy state (Miles *et al* 2001, Jayasooriya and Jenkins 2002). When molecules absorb energy, they are excited to a virtual energy state (figure 5.2). They promptly release this energy and settle into a lower vibrational state. As the molecules return to their ground state without gaining or losing any energy, the

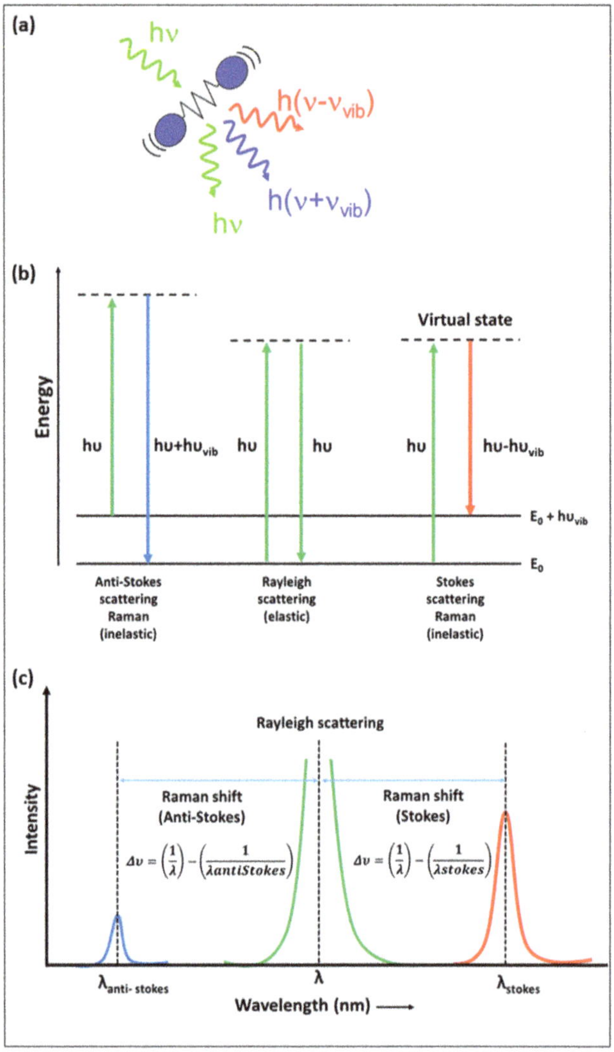

Figure 5.2. (a) Scattering of photons from a molecule. (b) Energy level diagram illustrating the Rayleigh and Raman scattering processes. (c) Raman shifts for Stokes and anti-Stokes scattering.

energy of the incident and scattered light remains the same. This process is referred to as elastic scattering. When the scattered light possesses less energy than the incident light, it is referred to as Stokes–Raman scattering. In contrast, if a molecule in a higher vibrational state moves to a virtual energy state and then relaxes to a lower vibrational level of the ground state, the scattered photons will have more energy than the incident photons. This process is known as anti-Stokes–Raman scattering. Usually, the Stokes–Raman scattering are routinely in use for the analysis due to the higher signal strength than the anti-Stokes–Raman scattering (Jayasooriya and Jenkins 2002).

In Raman scattering, the light photons cause a frequency shift after the interaction with the molecule, and this shift is plotted against the intensity of the scattered radiation, giving rise to the Raman spectrum, which reveals molecular vibrations linked to chemical bonds, providing insights into molecular structure, composition, and interactions. Tylenol (figure 5.3) is one of the standard molecules commonly used for the testing of Raman spectrometers due to its very intense Raman activity.

A typical Raman spectrum of the tylenol molecule is shown in figure 5.4. Each peak in the spectrum is attributed to the Raman shift corresponding to the vibrations of different functional groups and molecular bonds present in the molecule. The Raman band assignments of tylenol are also given in table 5.1.

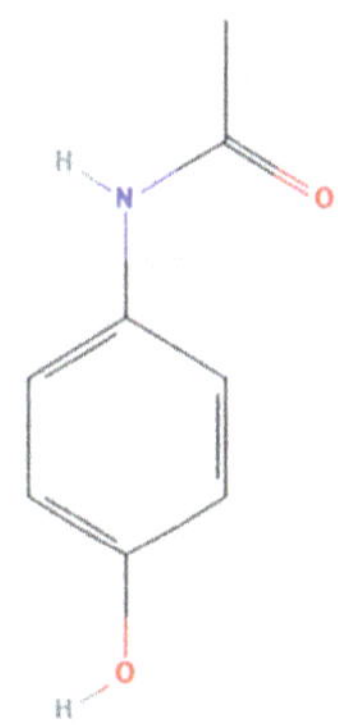

Figure 5.3. The chemical structure of a tylenol molecule (PubChem 2024).

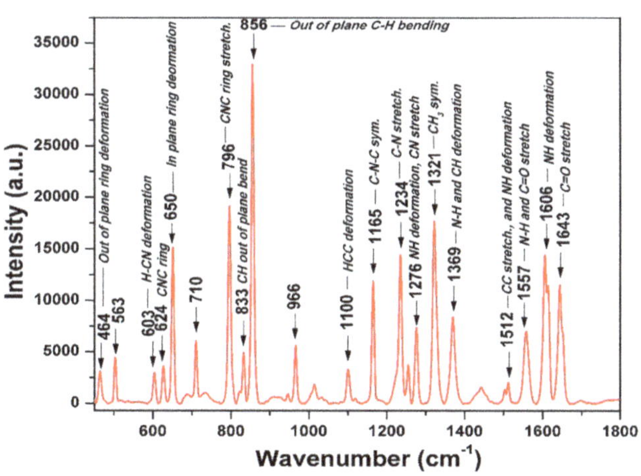

Figure 5.4. Typical Raman spectrum of tylenol.

Table 5.1. Band assignment of tylenol Raman spectrum.

Wavenumber (cm^{-1})	Band assignment
464	Out of plane ring deformation
603	H–CN deformation
624	CNC ring
650	In plane ring deformation
796	CNC ring stretch
833	CH out of plane bend
856	Out of plane CH bend
1100	HCC deformation
1165	C–N–C symmetric
1234	C–N stretch
1276	NH deformation, CN stretch
1321	CH_3 symmetric
1369	NH and CH deformation
1512	CC stretch, and NH deformation
1557	N–H and C=O stretch
1606	NH deformation
1643	C=O stretch

5.1.1.4 Raman spectroscopy advantages for biological studies

Raman spectroscopy is largely unaffected by water because water has weak Raman signals. This makes it ideal for studying biological samples in aqueous environments, which is crucial because most biological systems, like cells and tissues, exist in water-based environments. Raman measurements can be performed on biological samples with little to no preparation. Most of the samples can be studied in their native state without drying or special treatments, preserving the biological integrity of the sample. Raman spectroscopy provides detailed information on a wide range of biological molecules, including proteins, lipids, nucleic acids (DNA and RNA), and carbohydrates. It can reveal vibrational modes from the backbone and side chains of proteins, DNA bases, lipid structures, etc. Raman spectroscopy is particularly useful for studying protein secondary structures and conformations. Raman spectroscopy is complementary to fluorescence techniques and can provide additional molecular information where fluorescence is not applicable or desirable. It can also avoid fluorescence quenching or interference. Raman spectroscopy is not highly sensitive to the thickness of the sample, which allows for the analysis of relatively thick biological samples, such as tissues and biofilms.

Raman spectroscopy allows for label-free imaging of biological samples. Techniques such as coherent anti-Stokes–Raman scattering (CARS) and stimulated Raman scattering (SRS) enable high-speed, high-resolution, label-free imaging of live cells and tissues. These methods provide vibrational contrast without fluorescent or chemical labels, avoiding potential interference or toxicity.

Raman spectroscopy provides detailed molecular vibrational information, which can be directly correlated with the chemical structure, conformation, and functional groups of biomolecules. This can be used to investigate biomolecular changes in processes like protein folding, lipid organization, or disease progression. Resonance Raman spectroscopy can enhance the Raman signal of certain biomolecules (like heme-containing proteins, chromophores, etc) by tuning the excitation wavelength to match the electronic transition of the molecule. This can provide highly selective information about specific biological molecules, even at low concentrations.

5.1.2 Micro-Raman spectroscopy

It is a specialized technique used to analyze the chemical composition, molecular structure, crystalline properties of materials and biological samples on a microscopic scale. This method uses a Raman spectrometer integrated with a microscope to focus a laser beam onto a very small region of a sample, sometimes just a few microns in diameter. Micro-Raman technique offers high spatial resolution and can be used to map molecular distributions in tissues and cells. Raman microscopy can achieve sub-micron resolution, making it ideal for studying cellular structures, organelles, or even single cells.

5.1.2.1 Experimental setup to record micro-Raman spectra

An illustration of a Raman instrument is shown in figure 5.5. The laser source of the desired wavelength will excite the sample of interest. The sample for the measurement can be chosen using a microscope objective to focus on the micron-sized area of any sample. Hence, the name of the Raman spectrometer becomes micro-Raman.

A beam expander and a dichroic mirror are utilized to direct the expanded laser beam, ensuring it overfills the microscope objective for precise light focusing and permits the backscattered light signal to reach the Raman spectrometer. The Rayleigh scattered light is eliminated using an edge filter (or a notch filter) placed in front of the spectrometer, allowing the Raman signal to reach the spectrometer.

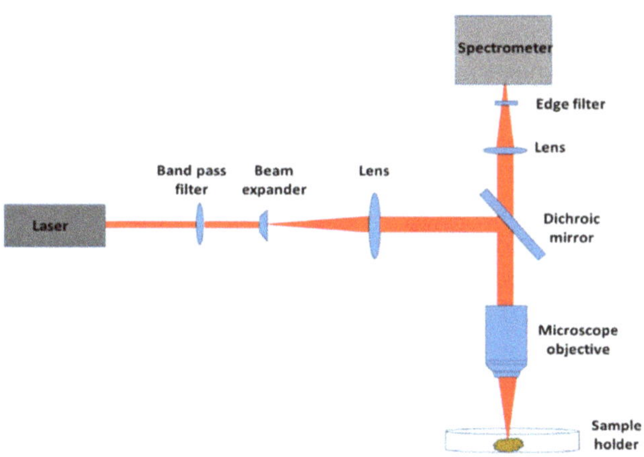

Figure 5.5. Illustration of a micro-Raman setup.

5.1.2.2 Medical application of micro-Raman spectroscopy

Raman spectroscopy is a highly versatile tool for analysing biological samples where water is a significant component. Analysing blood plasma and serum using this technique yields crucial insights into the details of proteins, lipids, and other biomolecules in the bloodstream. Raman spectroscopy-based analyses of biological samples typically utilise excitation wavelengths in the near-infrared region, usually 785 nm wavelength. This choice primarily aims to suppress fluorescence background interference and minimise sample damage caused by absorption effects.

5.1.2.3 Blood plasma and serum analysis

Blood constitutes plasma, red blood cells (RBCs), platelets, white blood cells (WBCs), proteins, amino acids, lipids, water, electrolytes, etc. Plasma is the major component and most significant part of blood. Plasma constitutes over half of the blood's total content, accounting for about 55%. When isolated from the rest of the blood, it appears as a pale-yellow liquid. Its primary function is transporting nutrients, proteins, and hormones to several organs of the human body. In addition, cells expel their waste products into the plasma, which aids in eliminating these materials from the body. Furthermore, blood plasma facilitates the circulation of blood components throughout the circulatory system. Blood plasma is separated from whole blood through a process called blood fractionation, using an anti-coagulant in a centrifuge to allow the blood cells to settle at the bottom of the vial. The plasma is then extracted for analysis. Blood serum is blood plasma without clotting factors.

5.1.2.4 Micro-Raman spectroscopy analysis of blood plasma and blood serum for the detection of cancers

When a healthy person becomes unwell, discernible alterations occur in their blood and its components. The alterations will be particularly reflected in blood plasma/blood serum. Analysing blood plasma/blood serum will yield comprehensive insights into the various disease states. Both blood plasma and serum contain numerous biomarkers indicative of different disease conditions. The various stages of cancer can be distinguished using Raman spectroscopy. This technique, combined with principal component analysis (PCA), can classify breast cancer in comparison to healthy controls, as well as differentiate between the different stages of the disease (Nargis *et al* 2019). Figure 5.6 shows the Raman spectra of the plasma samples of breast cancer patients and healthy individuals.

Compared to the plasma of cancer patients, the normal patient's plasma samples have higher Raman band intensities corresponding to nucleotides (689 cm^{-1}), phosphate (770 cm^{-1}), tyrosine (828 cm^{-1}), phospholipids (1268 cm^{-1}), and others (Krishnamoorthy *et al* 2019). The Raman bands of the post-treated samples show similarities to those of healthy samples. Additionally, Raman bands related to collagen, nucleic acids, and the DNA base pair adenine–guanine show variations between cancerous and normal conditions, as well as between pre-treatment and post-treatment cancer states.

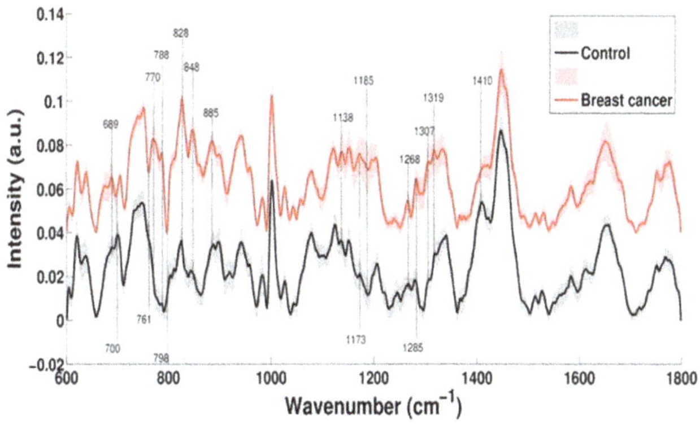

Figure 5.6. Raman spectra of blood plasma collected from healthy individuals and breast cancer patients (Nargis *et al* 2019).

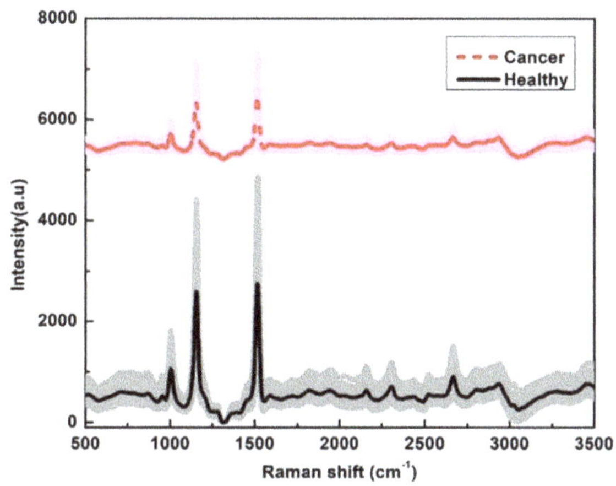

Figure 5.7. The average Raman spectra of serum samples collected from healthy volunteers and volunteers with gastric cancer (Bahreini *et al* 2019).

The Raman bands corresponding to cholesterol and proteins in serum samples have changes in gastric cancer (Bahreini *et al* 2019). Figure 5.7 illustrates the Raman spectra of a serum sample from gastric cancer patients compared to healthy individuals. Gastric cancer patients generally display decreased albumin levels in their blood serum. Moreover, the Raman bands corresponding to phenylalanine at 1000 cm^{-1} show reduced intensity in the serum of these patients (Li *et al* 2021). Furthermore, the beta-carotene bands at 1152 and 1514 cm^{-1} also show reduced intensity in these patients. Conversely, there is an increased intensity of CH_2 bending in collagen and phospholipids in individuals with gastric cancer. The Raman spectra of blood plasma from gastric cancer patients exhibit an extra band at 1182 cm^{-1}. Moreover, there is an increased Raman intensity associated with the amide III, II, and I regions of proteins, along with CH_2 groups and lipids (Guleken *et al* 2023). The characteristic Raman markers for gastric cancer were identified at 1302 and 1306

cm^{-1}. Extensive studies are carried out on oral, cervical, bladder, and other cancers using Raman spectroscopy techniques (Auner *et al* 2018, Falamas *et al* 2021).

5.2 Surface-enhanced Raman spectroscopy (SERS)

Usually, Raman signals are quite weak, making it challenging to study biological systems, and the quantity of samples is always minimal for measurement. To overcome this obstacle, a highly sensitive technique involves enhancing weak Raman scattered signals to several orders of magnitude using nanostructured materials. Surface-enhanced Raman spectroscopy (SERS) facilitates the characterization of low-concentration analytes via plasmon-mediated processes (Han *et al* 2021). Figure 5.8 shows the schematic diagram of the metallic nanoparticles-based SERS substrate.

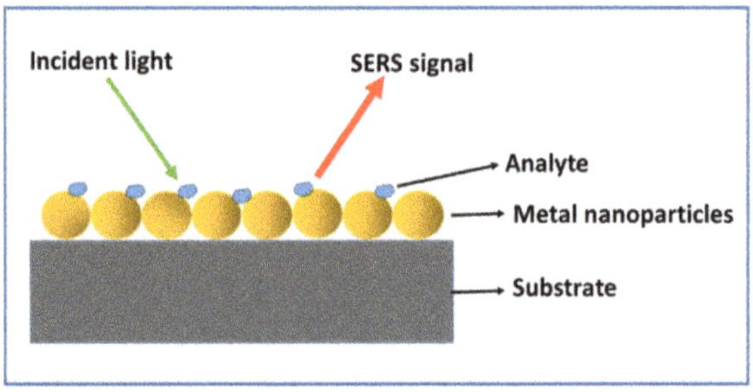

Figure 5.8. Schematic representation of nanoparticle-based SERS substrate.

5.2.1 Fundamentals of SERS

SERS is attributed to two mechanisms: the electromagnetic enhancement mechanism and the chemical (molecular) enhancement mechanism (Prochazka 2016). Usually, the light scattering can be related to the polarizability of molecules that interact with the electromagnetic field. The Raman effect can be described by the magnitude of the induced dipole moment (μ_{ind}), which is the product of the 'molecular polarizability (α_{molecule}), and the local electric field (E_{loc}), as expressed in equation (5.1). When molecules interact with metal surfaces, this interaction enhances either the local electric field or the molecular polarizability.

5.2.1.1 Electromagnetic enhancement

The main enhancement in SERS depends on increasing the electromagnetic (EM) field through the resonance excitations of localized conduction-electron oscillations at the surfaces of metallic nanostructures (Prochazka 2016, Bousiakou *et al* 2019). The resonance frequency (ω_{max}) of plasmons within the metallic nanostructure is determined by the dielectric function of the metal, represented as $\epsilon_{\text{metal}}(\omega)$, and the dielectric function of the surrounding medium, denoted as $\epsilon_m(\omega)$. At first, the interaction between the photon and localized surface plasmons (LSPs) creates a coupled state, resulting in a greatly enhanced EM field amplitude near the rough surface of the metal.

The molecule attached to the surface experiences a markedly stronger local electric field (E_{loc}). Secondly, the molecular dipole produces Raman scattering (RS) not in free space, but instead near the metal surface. The frequency-shifted Stokes–RS radiation at $\omega_s = \omega_{\mathrm{inc}} - \omega_{vib}$ for a particular vibrational mode can induce a Localized surface plasmon resonance (LSPR) in the metallic nanostructure. Here, ω_{inc} represents the incident frequency and ω_{vib} denotes the scattering frequency. The LSPR, coupled with the electromagnetic field, is the same physical phenomenon responsible for both local field enhancement and Raman radiation enhancement. The intensity of the SERS signal depends on incoming (ω_{inc}) and outgoing ($\omega_s = \omega_{\mathrm{inc}} - \omega_{vib}$) field.

$$I_{\mathrm{SERS}} = I_{inc}(\omega_{inc})I(\omega_s) = |E_{inc}(\omega_{inc})|^2 |E(\omega_s)|^2 \tag{5.7}$$

The required condition for SERS is both incident radiation at ω_{inc} and $\omega_s = \omega_{\mathrm{inc}} - \omega_{vib}$ should be in resonance with LSPR peak of the metallic nanostructure. The slight increase of $E_{\mathrm{loc}}/E_{\mathrm{inc}}$ will lead to a large enhancement of Raman scattering, which is termed an enhancement factor (EF). Usually, it can be of the order of 10^8. The electromagnetic enhancement is also affected by the shape and size of the metallic nanostructure, along with the distance separating the molecule from the metallic nanostructure. The enhancement factor for a small metallic sphere with a radius r, which is one-twentieth of the wavelength of the incident light, can be expressed as follows:

$$EF_{\mathrm{em}}(\omega_s) = \left| \frac{\epsilon(\omega_{\mathrm{inc}}) - \epsilon_0}{\epsilon(\omega_{\mathrm{inc}}) + 2\epsilon_0} \right|^2 \left| \frac{\epsilon(\omega_s) - \epsilon_0}{\epsilon(\omega_s) + 2\epsilon_0} \right|^2 (r/(r + d)^{12}) \tag{5.8}$$

where $\epsilon(\omega)$ is a complex frequency-dependent dielectric function of the metal, ϵ_0 is dielectric constant of the bulk medium, and d is distance of the molecule from the surface.

The intensity of the Raman band is influenced by the molecule's location relative to the surface. In SERS, substantial enhancement occurs when the real part of $\varepsilon(\omega)$ at the incident and/or scattered wavelengths is nearly equal to $-2\varepsilon_0$, with the imaginary part of $\varepsilon(\omega)$ being minimal.

5.2.1.2 Chemical enhancement

The molecular mechanism entails an increase in the molecular polarizability tensor ($\alpha_{\mathrm{molecule}}$), leading to an enhancement of the Raman cross-section (Xia *et al* 2014). This mechanism requires certain conditions, such as the existence of specific active sites and the formation of a metal–adsorbate chemical bond, commonly referred to as the first-layer effect (Prochazka 2016). The chemical enhancement mechanism includes the formation of charge-transfer complexes or alterations in molecular electronic states resulting from the interaction between the adsorbed molecule and the metal surface. This interaction alters the molecular polarizability and changes the 'Raman scattering cross-section' of the molecule, resulting in additional enhancement of the Raman signal. Typically, the chemical mechanism in SERS is less efficient compared to the EM mechanism. Experimental findings and theoretical assessments indicate EFs ranging from 10^0 to 10^2, contingent upon factors such as chemical structure, metal interaction, and specific vibrational mode (Valley *et al* 2013). Chemical enhancement tends to be more molecule-specific and can provide additional enhancement besides the electromagnetic enhancement.

Both EM and chemical mechanisms result in the overall large enhancement factor in SERS, which causes the highly sensitive technique for detecting very low concentrations of molecules, in some cases reaching a single molecular level.

5.2.2 Types of SERS substrates

To achieve sensitive SERS detection, developing a substrate with high signal enhancement, excellent repeatability, and better stability is essential. Both label-free and label-based SERS substrates can be employed to obtain the fingerprint information of analyte molecules (Zheng *et al* 2018, Chen *et al* 2023). Label-based SERS substrates utilise a Raman-active label (or tag) that attaches to the analyte molecule. The label has a distinct Raman signal, which is enhanced by the substrate. On the other hand, label-free SERS substrates do not require an external label, the Raman signal is directly obtained from the target molecule itself, which is enhanced by the substrate (Zheng *et al* 2018).

Label-free SERS substrates have garnered significant attention because of their simple preparation procedures. In addition, label-free detection will provide quantitative and qualitative information of several analytes. They can also directly deliver the intrinsic SERS signal of the analyte molecule within a complex biological sample, without labelling or any special pre-treatment. Label-free SERS substrate can be employed if the analyte molecule has a very high Raman cross-section. Noble metals, inorganic and organic materials are commonly used to fabricate SERS substrates (Demirel *et al* 2018, Chen *et al* 2023). Noble metals, such as silver (Ag) and gold (Au) exhibit strong LSPR, which significantly enhance the electromagnetic fields at their surfaces. The spatial regions at the metallic surface with extremely high local fields relative to the excitation field are called hot spots.

SERS hotspots can be generated by controlling the shape of metallic nanoparticles and designing nanostructures with narrow gaps or sharp extensions, such as bipyramidal nanoparticles, nanocubes, nanostars, nano-pine trees, nanorods, and core-shell structures (McLellan *et al* 2007, El-Said *et al* 2016, He *et al* 2018, Yao *et al* 2022). The SERS activities of different shapes of nanoparticles (truncated nanocubes, nanocubes with sharp corners, and bipyramid nanoparticles) were compared in figure 5.9. The figure shows that truncated nanocubes had a greater Raman intensity, the bipyramid nanoparticles and the nanocubes with sharp corners exhibited identical Raman intensities (McLellan *et al* 2007).

Another common method for creating SERS substrates using colloidal metallic nanoparticles, mainly gold (Au) and silver (Ag), dispersed in an aqueous solution. These nanoparticles display distinctive optical properties because of their LSPR, which intensifies the electromagnetic field in their vicinity when exposed to laser

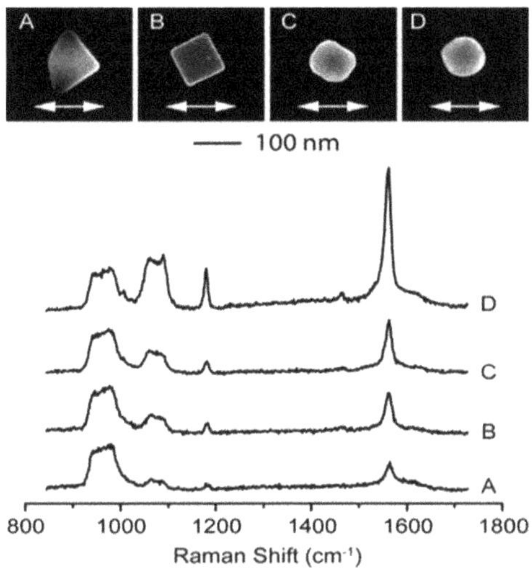

Figure 5.9. The SERS spectra of 1,4-benzenedithiol (1,4-BDT) adsorbed on silver (Ag) nanoparticles with various shapes: (A) right bipyramid, (B) cube with sharp corners, and (C and D) truncated cubes, as illustrated in the corresponding SEM images (McLellan *et al* 2007).

light. The preparation of colloidal SERS substrates typically involves synthesizing nanoparticles through methods such as seed-mediated growth or chemical reduction. The size, shape, and distribution of these nanoparticles can significantly influence the SERS enhancement factor, allowing for tunable sensitivity. Additionally, these colloidal substrates can be easily functionalized with specific molecules to target a particular analyte, enhancing selectivity in sensing applications.

Inorganic SERS substrates are mainly based on semiconductor and lamellar materials (Chen *et al* 2023). The chemical enhancement mechanism has an important role in the design of semiconducting SERS substrates. Semiconductors like ZnO, WO_3 and TiO_2 are commonly used to fabricate SERS substrates because of their high stability, low cost, and outstanding optical performance (Zheng *et al* 2017). Lamellar materials that are frequently employed as SERS substrates include graphene and MXene (Wang *et al* 2015, Jebakumari *et al* 2023).

SERS tags are present on nearly all labelled SERS substrates (Wang *et al* 2022). SERS tags and label-based SERS substrates have two different kinds of connections. The first method consists of combining the SERS tags directly with the analyte molecules to form a colloidal SERS substrate, which is subsequently utilized for SERS detection. The second approach involves attaching SERS tags to different substrate materials to fabricate solid SERS substrates for detection. SERS tags offer several advantages, including the ability for single-particle sensing, reusability, and stability. As a result, SERS technology is anticipated to take the place of conventional optical tags, like quantum dots and fluorescent dyes, in a broad array of biosensor applications (Chen *et al* 2023). SERS tags can provide detection information in three different ways: (1) SERS tags act as internal tracers by combining with the target molecule. (2) The SERS tag competes with the target

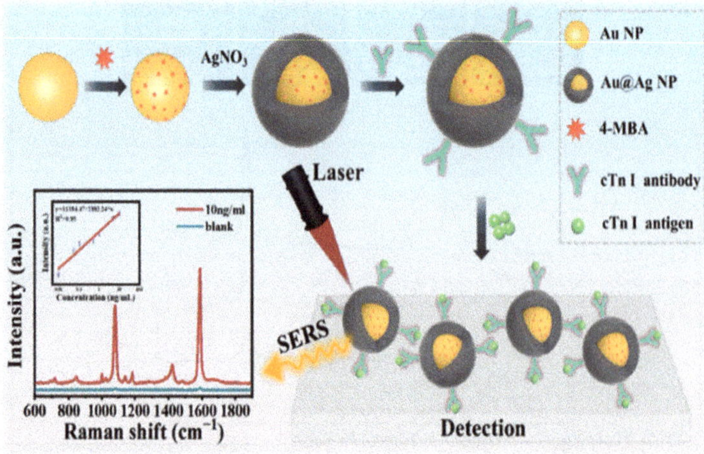

Figure 5.10. Schematic representation of the preparation of SERS nanotags and the detection of cardiac troponin I (cTn I) (Wang *et al* 2022).

molecule for substrate adsorption. (3) The SERS tag interacts with the target molecule and causes changes in the Raman spectrum (Chen *et al* 2023). Figure 5.10 depicts the preparation of SERS nanotags by embedding 4-mercaptobenzoic acid (4-MBA) between bimetallic gold (Au) and silver (Ag) nanospheres, resulting in the formation of Au@4-MBA@Ag. This configuration was then coupled with an antibody to create an immunological probe used for the detection of cardiac troponin I (cTn I).

5.2.3 Medical applications of the SERS technique

The SERS technique can enhance the Raman signal strength for biomarker detection, thereby disease diagnosis. A few examples are discussed here. In a research study, gold nanoparticles (Au–NPs) were effectively used as a substrate to enhance SERS spectra from blood plasma. This study involved 160 subjects, with 60 histopathologically normal and 100 diagnosed with nasopharyngeal carcinoma (NPC). Among the NPC cases, 25 were classified as T1 stage and 75 spanned stages T2 to T4. The use of Au–NPs significantly boosted Raman signal intensities, as shown in figure 5.11, where vibrational band intensity in plasma on Au–NPs was markedly higher than that from traditional Raman spectroscopy. This enhancement is attributed to the LSPR effect of Au–NPs interacting with plasma biomolecules. These findings underscore the potential of Au–NP-based SERS as a non-invasive, sensitive diagnostic tool for cancer detection and staging, promising improved early screening for NPC in clinical applications (Vargas-Obieta *et al* 2016).

Another study performed SERS measurements on two groups of blood serum samples: one from patients with confirmed colorectal cancer and the other from healthy volunteers. Significant differences in Raman spectra have been observed between normal and cancerous serum samples. The normalized intensities of SERS

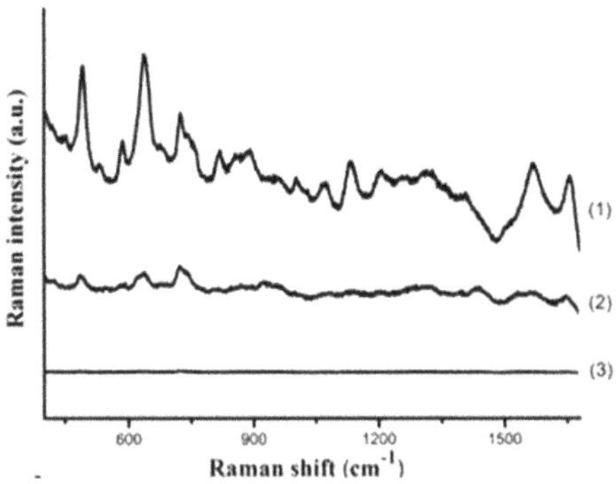

Figure 5.11. Comparison of (1) SERS spectrum of blood plasma-Au NPs mixture, (2) regular Raman spectrum of the same plasma sample without the Au NPs and (3) background Raman signal of the anticoagulant mixed with Au colloid (Vargas-Obieta *et al* 2016).

peaks at 494, 638, 823, 1206, and 1655 cm^{-1} were lower in cancer samples compared to normal samples. On the other hand, the SERS bands at 725 and 881 cm^{-1} showed higher intensities in cancer samples. The SERS spectra of human serum in the 350–1700 cm^{-1} range are primarily characterized by the vibrational modes of biomolecules like nucleic acids, proteins, and lipids. These biomolecules may undergo changes in their quantity or conformation, which are associated with the progression of colorectal cancer (Lin *et al* 2011). Based on blood plasma SERS analysis, cervical cancer detection is presented (Feng *et al* 2013). A comparison of the SERS spectra between cervical cancer plasma and normal plasma reveals significant differences in spectral intensities. Specifically, cancer plasma exhibits lower intensities at 496, 534, 638, 813, 888, 1135, and 1400 cm^{-1}, while higher intensities are observed at 1578 and 1655 cm^{-1}. These spectral variations highlight the potential of SERS as an effective diagnostic tool for non-invasive detection of cervical cancer.

5.3 Raman tweezers spectroscopy

Raman tweezers is a combination of two powerful techniques: Raman spectroscopy and optical tweezers. This hybrid method allows for the manipulation and analysis of microscopic particles, such as biological cells or molecules, using focused laser beams. The technique of optical tweezers was demonstrated in 1984 at Bell Laboratories. In 1987, they showed the damage-free manipulation of cells using an infrared laser. Arthur Ashkin got the Nobel Prize in 2018 for this invention.

5.3.1 Fundamentals of optical tweezers

Optical tweezers can manipulate dielectric particles ranging from nanometers to microns in size by applying minute forces using a tightly focused laser beam (figure 5.12). The forces responsible for optical trapping are gradient force and scattering force. The gradient force pulls the particle toward the region of highest light intensity (the focal point of the laser beam). It is responsible for holding the

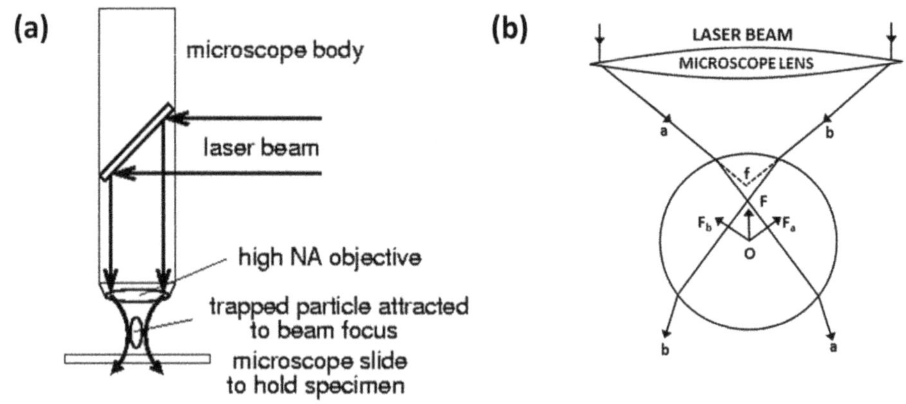

Figure 5.12. (a) Illustration of optical trapping. (b) Ray diagram for the trapping of a dielectric sphere.

particle in place, as the particle experiences a restoring force when it is displaced from the center of the trap. The scattering force is caused by the momentum transfer from the photons to the particle as light is scattered. It pushes the particle along the direction of the laser beam (Prasad 2004).

When dealing with particles significantly larger than the wavelength of the incident laser, known as the Mie regime, trapping can be described using ray optics. Take, for instance, a dielectric sphere with a refractive index higher than that of the surrounding medium. As light passes through the sphere, it bends due to refraction. The principle of momentum conservation dictates that the change in momentum of the refracted rays results in an equal and opposite transfer of momentum to the sphere. A particle with a higher refractive index than the surrounding medium will be pulled toward the region of maximum intensity (the laser focus). This force helps in trapping particles in three dimensions. This force is proportional to the electric field gradient and the polarizability of the particle.

The mathematical expression of gradient force can be written as

$$F_{\text{gradient}} \propto \alpha \, \nabla \, (|E|^2) \tag{5.9}$$

where α is polarizability of the particle and E is electric field of the laser beam.

Therefore, a focused laser can establish a trap because the light intensity gradient is directed toward the centre. However, if the gradient force is not stronger than the scattering force, the particle may escape along the optical axis. The light gradient must be steep to achieve a stable trap, which can be accomplished using a high numerical aperture microscope objective (Prasad 2004).

The idea of optical trapping employing two rays of a focusing beam (labelled as (a) and (b)) is illustrated in figure 5.12(b), assuming no surface reflection. Fa and Fb represent the forces resulting from beam refraction and are aligned with the direction of momentum change. As depicted in figure 5.12(b), it's clear that the restoring force is directed towards the focus regardless of the dielectric sphere's centre position.

5.3.2 Experimental setup for single live cell Raman spectroscopy

The experimental setup for trapping micron-sized particles is illustrated in figure 5.13, while figure 5.14 displays the recorded Raman spectrum of an optically trapped polystyrene bead. The band assignments are also indicated on the graph. Figure 5.13 shows the schematic diagram of an optical tweezers Raman spectroscopy system. It has a 785 nm diode laser for trapping and probing purposes. The holographic bandpass filter filters out unwanted spurious lines from the laser. A 1:1 telescopic arrangement is incorporated to manipulate laser spots in the sample

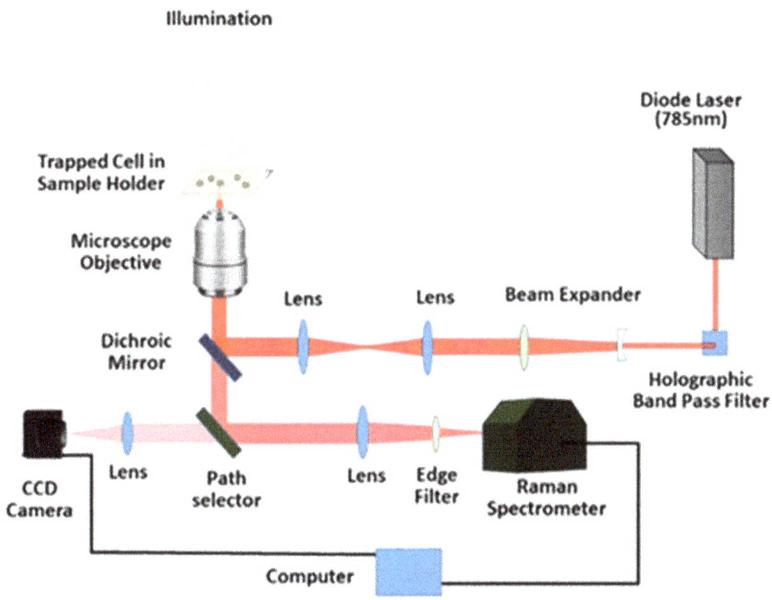

Figure 5.13. Schematics of Raman tweezers spectroscopy setup (Lukose *et al* 2020).

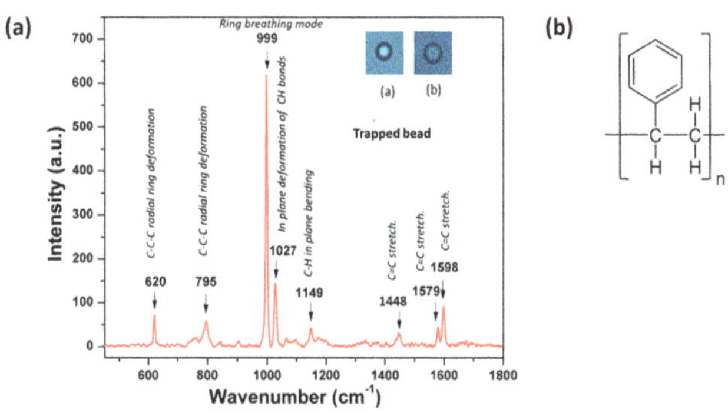

Figure 5.14. (a) Raman spectrum of an optically trapped polystyrene bead and (b) chemical structure of polystyrene.

plane. The dichroic mirror in the setup reflects the 785 nm laser beam and transmits the Stokes–Raman signal. An inverted microscope is incorporated into the setup to enable the laser beam to focus from the bottom of the sample holder to create the laser trap. The 100× microscope objective is utilised to direct light onto the sample and collect backscattered Raman signal. The scattered signal passes through the dichroic mirror and reaches an optical path selector that directs the signal to the microscope camera to see the trapped object and to the spectrometer to record Raman spectra. The scattered light from the sample can be filtered using an edge filter to separate out the Raman scattered signal.

The standardization and calibration of a Raman tweezers setup can be performed by optically trapping and recording the Raman spectrum of polystyrene beads. A typical Raman spectrum of a trapped polystyrene bead is displayed in figure 5.14.

5.3.3 Medical application of Raman tweezers spectroscopy

This technique allows for the precise manipulation and analysis of biological samples at the single-cell level. By trapping cells with laser light and obtaining molecular information through Raman scattering, researchers can study cellular composition, detect diseases, and monitor drug interactions in real time. Applications include monitoring cellular responses to external stressors and therapies, cell–cell interactions, and understanding disease mechanisms at a molecular level.

5.3.3.1 Raman spectra of a single live red blood cell

Raman tweezers systems have been extensively employed to analyse red blood cells (RBCs), white blood cells (WBCs), and platelets and their interaction with external stressors (Barkur *et al* 2017). Figure 5.15 presents the Raman spectra for single live RBC. The observed Raman bands in RBCs primarily originate from proteins and lipids. Figure 5.15 shows the Raman spectrum single live RBC, with the microscopic images (given in the inset) of optically trapped RBC (flipped RBC) and floating RBC (discoid shape).

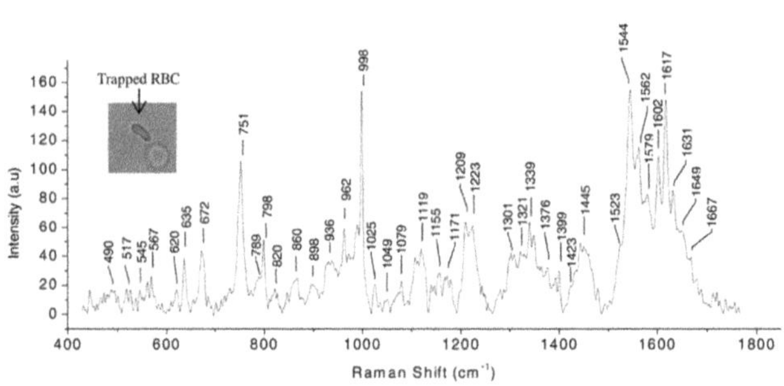

Figure 5.15. Raman spectra of red blood cell (Bankapur *et al* 2010).

5.3.3.2 Raman tweezers spectroscopy of blood cells interaction with bisphenol-A

The impact of bisphenol A (2,2Di(p-hydroxyphenyl)propane, BPA) on RBCs was examined using the Raman tweezers technique (Lukose *et al* 2019). The Raman spectra of bisphenol A treated RBCs were shown in figure 5.16. The increased BPA level in blood have been linked to various health problems, including reproductive disorders, cardiovascular diseases, diabetes, autism, and obesity. The reduced intensities at 752 cm^{-1} (porphyrin breathing mode) and 999 cm^{-1} suggest depletion of haemoglobin and RBC membranes. Additionally, the transition from deoxygenated to oxygenated states of RBCs is indicated by the frequency shift from 1209 cm^{-1} (C–H bending) to 1222 cm^{-1}.

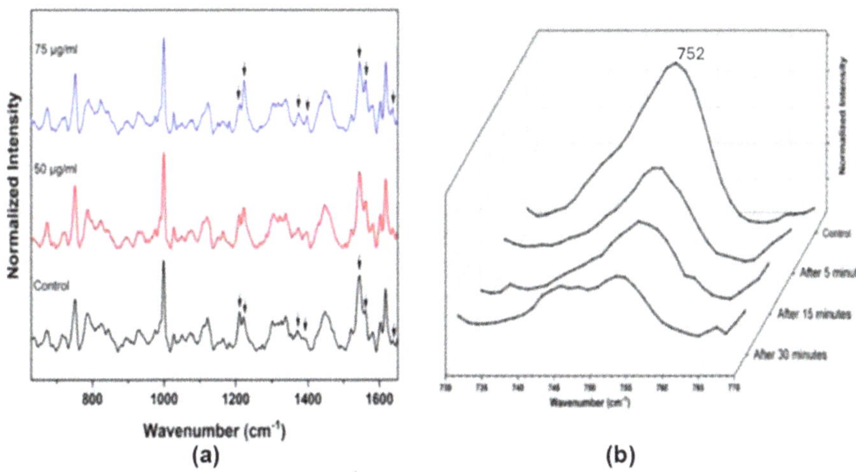

Figure 5.16. Raman spectra of (a) control RBCs and RBCs treated with varying BPA concentrations. (b) Raman peak intensity of porphyrin breathing mode (Lukose *et al* 2019).

5.3.3.3 Raman tweezers spectroscopy of blood cells interaction with ethanol

The effect of ethanol on individual live RBCs is evidenced by the variation in intensity of specific Raman bands (Lukose *et al* 2019). The decreased intensity at 1222, 1561, and 1636 cm^{-1} indicates a deoxygenation in ethanol-treated RBCs. Additionally, the intensity of the porphyrin breathing mode at 752 cm^{-1} is also reduced. High concentrations of ethanol cause damage to the RBC membrane. Figure 5.17 illustrates the Raman spectra of RBCs exposed to different ethanol concentrations, with microscopic images of single RBCs treated with various ethanol levels and in control plasma shown in the inset. At higher ethanol concentrations, the RBCs' discoid shape transforms into echinocytes.

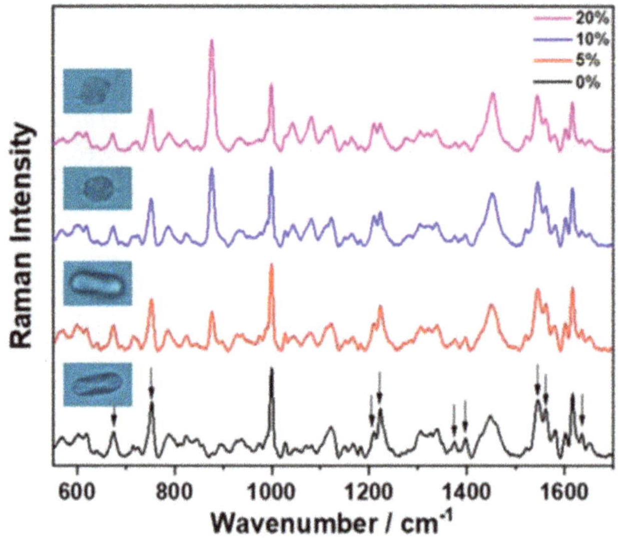

Figure 5.17. Raman spectra of RBCs treated with varying ethanol concentrations. The microscopic images were given in the inset (Lukose *et al* 2019).

5.3.3.4 Raman tweezers spectroscopy analysis of different cancer cells

Astrocytoma, a type of brain cancer cell, and astrocytes, a kind of normal cell, can be compared by recording their Raman spectra. Compared to normal astrocyte cells the astrocytoma cells have higher intensities for the peak at 878 cm^{-1}, which is from the tyrosine, 1004 cm^{-1} from phenylalanine, 1264 cm^{-1}, 1302 cm^{-1} from amide III, 1442 cm^{-1} from lipid, and 1660 cm^{-1} from amide I) (Banerjee and Zhang 2007, Banerjee *et al* 2015). The epithelial cancer investigation with the aid of Raman tweezers spectroscopy has provided remarkable results. The Raman band intensities of 788, 853, 938, 1004, 1095, 1257, 1304, 1446 and 1657 cm^{-1} correspond to DNA backbone OPO stretching, ring breathing mode of tyrosine, C–C backbone stretching of protein α helix, symmetric ring breathing of phenylalanine, DNA PO_2— symmetric stretching, amide III, lipids CH_2 twist, CH_2 deformation, and CO stretching of Amide I α helix, respectively, are higher in cancer cells. By comparing the data from Raman spectroscopy with medical findings, it was shown that malignant cells had a relatively greater level of nucleic acids and proteins than normal cells (Chen *et al* 2006). The diagnosis of colorectal cancer from the single live epithelial cells can also be investigated by Raman tweezers spectroscopy. The intensities of the Raman bands, particularly those associated with proteins and nucleic acids, are significantly higher, suggesting that cancer cells exhibit increased activities of DNA/RNA and proteins. Raman tweezers spectroscopy enables the differentiation between prostate cancer cells (PC-3) and bladder cancer cells (MGH-U1). The study's findings indicate higher levels of nucleic acids and proteins in MGH-U1 cells compared to PC-3, while PC-3 cells exhibited higher concentrations of lipids and carbohydrates (Harvey *et al* 2008). The leukaemia cells and normal

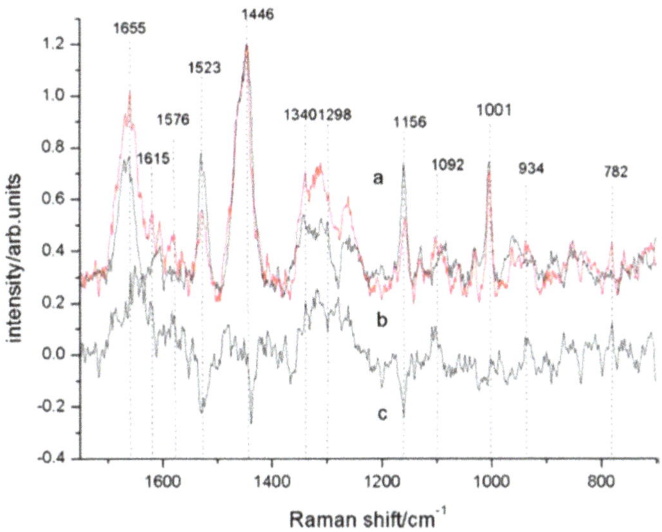

Figure 5.18. (a) The Raman spectrum of gastric carcinoma cells (without treatment). (b) Raman spectrum of apoptotic cells. (c) The difference spectrum of gastric carcinoma and apoptotic cells (Yao *et al* 2009).

lymphocytes can be investigated using the Raman tweezers spectroscopy method. There are specific Raman spectral features related to DNA, RNA, and protein molecular vibrations for differentiating patient-derived leukaemia cells from healthy human lymphocytes. Compared to normal cells, the Raman band intensity ratio of DNA to protein was lower in cancer cells (Chan *et al* 2008). The Raman spectra of gastric carcinoma cells are shown in figure 5.18 (Yao *et al* 2009). The remarkable spectral intensity variations for the Raman shifts are marked in the spectra.

The Raman tweezers spectroscopy study was performed on the peripheral blood of cancer patients, and the experiment was performed on healthy WBCs and RBCs and compared with the leukemic cells and the cells from solid tumours. According to the Raman spectra obtained from the experiment, utilizing support vector machines (SVMs), a non-linear pattern recognition algorithm, the normal cells and cancer cells are classified with >99.7% sensitivity and 99.5% specificity (Neugebauer *et al* 2010).

5.4 Conclusions

The three variants of Raman spectroscopy techniques, micro-Raman, surface-enhanced Raman and Raman's tweezers discussed here are popular in biomedical research for unravelling many biological problems. Micro-Raman spectrometers are widely used by academia and industry for material characterization and quality control of pharmaceutical products. Even though the Raman tweezers system become a routine technique used by the single-cell spectroscopy community, there is still no commercial instrument available for wide application. Among the three techniques discussed here, SERS is still a fast-developing research topic that pays more attention to fabricating highly stable, sensitive metal nanostructures with maximum enhancement factors.

5.5 Problems

1. A sample is illuminated with a laser light of wavelength 532 nm. The scattered light is measured, and a Raman peak is observed at a wavelength of 570 nm. Calculate the Raman shift in wavenumbers.
2. In a Raman spectrum, the intensity of the Stokes line is measured to be 10 000 counts, while the intensity of the anti-Stokes line is 4000 counts. Calculate the ratio of the Stokes to anti-Stokes intensities and interpret the significance of this ratio.
3. A Raman shift of 2100 cm^{-1} is observed for a particular vibrational mode of a molecule. Calculate the corresponding vibrational frequency in Hertz (Hz).
4. Given a Raman-active mode with a wavenumber shift of 200 cm^{-1} and a measured Stokes to anti-Stokes intensity ratio of 2, estimate the temperature of the sample.
5. The differential Raman scattering cross-section for a particular vibrational mode is given as 1018 $cm^2\ sr^{-1}$. If a laser with a power of 200 mW and a wavelength of 532 nm is used, calculate the number of scattered photons per second for a sample with a concentration of molecules/cm^3 in a detection solid angle of 1 sr.
6. A Raman spectrum of a molecule shows a Stokes line at 852 cm^{-1} and an anti-Stokes line at 818 cm^{-1}. Calculate the vibrational frequency of the molecule and determine the temperature at which the spectrum was recorded.
7. What is the frequency of the first anti-Stokes line if the wavelength of the incident radiation is 5×10^{-5} cm and the Stokes line is spaced 750 cm^{-1} from the Rayleigh line?
8. A Raman scattering experiment is performed on a benzene solution. The Raman scattering cross-section of benzene for the 992 cm^{-1} mode is 3×10^{-30} cm^2 $molecule^{-1}$. The laser power is 200 mW, focused to a spot size of 2 μm^2. The benzene concentration is 0.1 mol l^{-1}. The laser wavelength is 532 nm. Assuming 10% collection efficiency, calculate the Raman scattering signal (number of photons per second) for the 992 cm^{-1} mode.
9. A molecule exhibits a strong electronic absorption band at 300 nm. A Raman spectrum is recorded using a laser with a wavelength of 405 nm. Describe the expected impact on the Raman scattering intensity when compared to using a laser with a wavelength of 532 nm.
10. A Raman spectrum of a benzene solution is obtained using a laser with a wavelength of 532 nm. The intensity of the 992 cm^{-1} Raman band is measured for several solutions with different benzene concentrations. A calibration curve is constructed with the Raman band intensity on the *y*-axis and benzene concentration on the *x*-axis. The equation of the

calibration curve is found to be Intensity = 5000 × Concentration + 100. An unknown benzene solution is measured, and its Raman band intensity is found to be 35 100. What is the concentration of benzene in the unknown solution?

References

Auner G W, Koya S K, Huang C, Broadbent B, Trexler M, Auner Z, Elias A, Mehne K C and Brusatori M A 2018 Applications of Raman spectroscopy in cancer diagnosis *Cancer Metastasis Rev.* **37** 691–717

Bahreini M, Hosseinzadegan A, Rashidi A, Miri S R, Mirzaei H R and Hajian P 2019 A Raman-based serum constituents' analysis for gastric cancer diagnosis: *In vitro* study *Talanta* **204** 826–32

Banerjee H N and Zhang L 2007 Deciphering the finger prints of brain cancer astrocytoma in comparison to astrocytes by using near infrared Raman spectroscopy *Mol. Cell. Biochem.* **295** 237–40

Banerjee H N *et al* 2015 Deciphering the finger prints of brain cancer glioblastoma multiforme from four different patients by using near infrared Raman spectroscopy *J. Cancer Sci. Ther.* **7** 44

Bankapur A, Zachariah E, Chidangil S, Valiathan M and Mathur D 2010 Raman tweezers spectroscopy of live, single red and white blood cells *PLoS One* **5** e10427

Barkur S, Bankapur A, Chidangil S and Mathur D 2017 Effect of infrared light on live blood cells: role of β-carotene *J. Photochem. Photobiol.,* B **171** 104–16

Bousiakou L G, Gebavi H, Mikac L, Karapetis S and Ivanda M 2019 Surface enhanced Raman spectroscopy for molecular identification-A review on surface plasmon resonance (SPR) and localised surface plasmon resonance (LSPR) in optical nanobiosensing *Croat. Chem. Acta* **92** 479–94

Chan J W, Taylor D S, Lane S M, Zwerdling T, Tuscano J and Huser T 2008 Nondestructive identification of individual leukemia cells by laser trapping Raman spectroscopy *Anal. Chem.* **80** 2180–7

Chen K, Qin Y, Zheng F, Sun M and Shi D 2006 Diagnosis of colorectal cancer using Raman spectroscopy of laser-trapped single living epithelial cells *Opt. Lett.* **31** 2015–7

Chen Y, An Q, Teng K, Liu C, Sun F and Li G 2023 Application of SERS in in-vitro biomedical detection *Chem.– Asian J.* **18** e202201194

Demirel G, Usta H, Yilmaz M, Celik M, Alidagi H A and Buyukserin F 2018 Surface-enhanced Raman spectroscopy (SERS): an adventure from plasmonic metals to organic semiconductors as SERS platforms *J. Mater. Chem.* C **6** 5314–35

El-Said W A, Fouad D M and El-Safty S A 2016 Ultrasensitive label-free detection of cardiac biomarker myoglobin based on surface-enhanced Raman spectroscopy *Sens. Actuators* B **228** 401–9

Falamas A, Faur C I, Ciupe S, Chirila M, Rotaru H, Hedesiu M and Pinzaru S C 2021 Rapid and noninvasive diagnosis of oral and oropharyngeal cancer based on micro-Raman and FT-IR spectra of saliva *Spectrochim. Acta, Part* A **252** 119477

Feng S *et al* 2013 Blood plasma surface-enhanced Raman spectroscopy for non-invasive optical detection of cervical cancer *Analyst* **138** 3967–74

Guleken Z *et al* 2023 An application of raman spectroscopy in combination with machine learning to determine gastric cancer spectroscopy marker *Comput. Methods Programs Biomed.* **234** 107523

Han X X, Rodriguez R S, Haynes C L, Ozaki Y and Zhao B 2021 Surface-enhanced Raman spectroscopy *Nat. Rev. Methods Prim.* **1** 87

Harvey T J, Faria E C, Henderson A, Gazi E, Ward A D, Clarke N W, Brown M D, Snook R D and Gardner P 2008 Spectral discrimination of live prostate and bladder cancer cell lines using Raman optical tweezers *J. Biomed. Opt.* **13** 64004–4

He S, Kyaw Y M E, Tan E K M, Bekale L, Kang M W C, Kim S S Y, Tan I, Lam K P and Kah J C Y 2018 Quantitative and label-free detection of protein kinase A activity based on surface-enhanced raman spectroscopy with gold nanostars *Anal. Chem.* **90** 6071–80

Jayasooriya U A and Jenkins R D 2002 Introduction to Raman spectroscopy *An Introduction to Laser Spectroscopy* 2nd edn (Boston, MA: Springer) pp 77–104

Jebakumari K E, Murugasenapathi N K and Palanisamy T 2023 Engineered two-dimensional nanostructures as SERS substrates for biomolecule sensing: a review *Biosensors* **13** 102

Krishnamoorthy C, Prakasarao A, Srinivasan V, GN S P and Singaravelu G 2019 Monitoring of breast cancer patients under pre and post treated conditions using Raman spectroscopic analysis of blood plasma *Vib. Spectrosc.* **105** 102982

Li M, He H, Huang G, Lin B, Tian H, Xia K, Yuan C, Zhan X, Zhang Y and Fu W 2021 A novel and rapid serum detection technology for non-invasive screening of gastric cancer based on Raman spectroscopy combined with different machine learning methods *Front. Oncol.* **11** 665176

Lin D, Feng S, Pan J, Chen Y, Lin J, Chen G, Xie S, Zeng H and Chen R 2011 Colorectal cancer detection by gold nanoparticle based surface-enhanced Raman spectroscopy of blood serum and statistical analysis *Opt. Express* **19** 13565–77

Lukose J, Mithun N, Priyanka M, Mohan G, Shastry S and Chidangil S 2019 Laser Raman tweezer spectroscopy to explore the bisphenol A-induced changes in human erythrocytes *RSC Adv.* **9** 15933–40

Lukose J, N M, Mohan G, Shastry S and Chidangil S 2019 Optical tweezers combined with micro-Raman investigation of alcohol-induced changes on single, live red blood cells in blood plasma *J. Raman Spectrosc.* **50** 1367–74

Lukose J, Shastry S, Mithun N, Mohan G, Ahmed A and Chidangil S 2020 Red blood cells under varying extracellular tonicity conditions: an optical tweezers combined with micro-Raman study *Biomed. Phys. Eng. Express* **6** 015036

McLellan J M, Li Z Y, Siekkinen A R and Xia Y 2007 The SERS activity of a supported Ag nanocube strongly depends on its orientation relative to laser polarization *Nano Lett.* **7** 1013–7

Miles R B, Lempert W R and Forkey J N 2001 Laser rayleigh scattering *Meas. Sci. Technol.* **12** R33

Nargis H F *et al* 2019 Raman spectroscopy of blood plasma samples from breast cancer patients at different stages *Spectrochim. Acta, Part* A **222** 117210

Neugebauer U, Bocklitz T, Clement J H, Krafft C and Popp J 2010 Towards detection and identification of circulating tumour cells using Raman spectroscopy *Analyst* **135** 3178–82

Prasad P N 2004 *Introduction to Biophotonics* (New York: Wiley)

Prochazka M 2016 Surface-enhanced Raman spectroscopy *Biological and Medical Physics, Biomedical Engineering* (Springer) ch 6 127–48

Raman C V 1928 A new radiation *Indian J. Phys.* **2** 387–98

Valley N, Greeneltch N, Van Duyne R P and Schatz G C 2013 A look at the origin and magnitude of the chemical contribution to the enhancement mechanism of surface-enhanced Raman spectroscopy (SERS): theory and experiment *J. Phys. Chem. Lett.* **4** 2599–604

Vargas-Obieta E, Martínez-Espinosa J C, Martínez-Zerega B E, Jave-Suárez L F, Aguilar-Lemarroy A and González-Solís J L 2016 Breast cancer detection based on serum sample surface enhanced Raman spectroscopy *Lasers Med. Sci.* **31** 1317–24

Wang D, Zhao Y, Zhang S, Bao L, Li H, Xu J, He B and Hou X 2022 Reporter molecules embedded au@ ag core-shell nanospheres as SERS Nanotags for cardiac troponin I detection *Biosensors* **12** 1108

Wang P, Xia M, Liang O, Sun K, Cipriano A F, Schroeder T, Liu H and Xie Y H 2015 Label-free SERS selective detection of dopamine and serotonin using graphene-Au nanopyramid heterostructure *Anal. Chem.* **87** 10255–61

Xia L, Chen M, Zhao X, Zhang Z, Xia J, Xu H and Sun M 2014 Visualized method of chemical enhancement mechanism on SERS and TERS *J. Raman Spectrosc.* **45** 533–40

Yao H, Tao Z, Ai M, Peng L, Wang G, He B and Li Y Q 2009 Raman spectroscopic analysis of apoptosis of single human gastric cancer cells *Vib. Spectrosc.* **50** 193–7

Yao M M, Tang H, Yin Y C, Zhang X, Lu Y L, Zhao X X, Gan T and Xu W P 2022 Tuning the surface enhanced Raman spectroscopy performance of Au core-Ag shell nanostructure for label-free highly sensitive detection of colorectal cancer marker *J. Alloys Compd.* **896** 163043

Zheng X, Ren F, Zhang S, Zhang X, Wu H, Zhang X, Xing Z, Qin W, Liu Y and Jiang C 2017 A general method for large-scale fabrication of semiconducting oxides with high SERS sensitivity *ACS Appl. Mater. Interfaces* **9** 14534–44

Zheng X S, Jahn I J, Webe K, Cialla-May D and Popp J 2018 Label-free SERS in biological and biomedical applications: recent progress, current challenges and opportunities *Spectrochim. Acta Part* A **197** 56–77

IOP Publishing

Diagnostic Biomedical Optics
Fundamentals and applications
Murukeshan Vadakke Matham, C S Suchand Sandeep, Maria Merin Antony, Manojit Pramanik and Santhosh Chidangil

Chapter 6

Photoacoustic imaging for biomedical applications

Katherine Gisi, Vijitha Periyasamy, Avishek Das and Manojit Pramanik

In this chapter, we will introduce an emerging hybrid imaging modality known as photoacoustic imaging (PAI) for pre-clinical and clinical applications. A brief description of and rationale for PAI will be given. The different types of clinical/pre-clinical PAI systems will be described along with their respective strengths and weaknesses. The types of clinical systems discussed are categorized into whether they perform tomography, microscopy, and endoscopy. The chapter mainly focuses on recent applications of PAI categorized by human organs from a head-to-toe fashion and concludes with a look at surgical and monitoring devices based on PAI.

6.1 Introduction to photoacoustic imaging

Biomedical imaging is an essential component of disease diagnosis and treatment (Acharya *et al* 1995, Periyasamy *et al* 2024). In modern medicine and the age of big data, frequent monitoring of health parameters is a rapidly growing evolving field. Imaging applications encompass the smallest organelle to the largest organs to gather diverse information. Various energy sources and data acquisition methods contribute to the significant variability in information obtained from different tissue regions. Much of the electromagnetic spectrum (such as x-rays, gamma rays, radio waves, and light) and sound waves (ultrasound, or US) are utilized in biomedical imaging. X-rays (utilized in x-ray projection imaging and x-ray computed tomography or x-ray CT) and gamma rays (used in nuclear imaging techniques like single photon emission computed tomography (SPECT) or positron emission tomography (PET)) are forms of ionizing radiation. Meanwhile, radio frequency (RF) waves used in magnetic resonance imaging (MRI) are considered safe. However, MRI remains one of the costliest imaging methods. US imaging is affordable, portable, and real-time, although it offers limited soft tissue contrast and has operator-dependent image quality.

doi:10.1088/978-0-7503-2364-2ch6

Therefore, scientists are always looking for new imaging technology that can provide complementary information or added advantage over the existing ones.

Optical imaging (Zhu *et al* 2013, Kenry *et al* 2018, Dang *et al* 2019), for instance, delivers high-resolution images but faces challenges in imaging beyond superficial skin layers due to light scattering within biological tissue. Techniques like fluorescence microscopy (FM) and laser spectral imaging (LSI) offer approximately 1 μm resolution but are limited to an imaging depth of ~1 mm. Near infrared spectroscopy (NIRS), a diffused optical imaging method, achieves deeper imaging depths (up to 80 mm) but with lower spatial resolution (5–10 mm) (Dang *et al* 2019). On the other hand, US imaging can penetrate up to several tens of centimeters but sacrifices spatial resolution (1 mm) and soft tissue contrast (Emilio 2019). Combining optics and US through photoacoustics presents an opportunity to overcome the limitations of both optical and US imaging.

Photoacoustic imaging (PAI) harnesses the photoacoustic (PA) effect, first observed by Alexander Graham Bell in 1880 during a photophone experiment (Manohar and Razansky 2016). He observed that sound waves were generated upon absorption of modulated light by a metal sheet. Initially used mostly in gas analysis, its applications expanded to solids and liquids in the 1970s, focusing on surface heating effects. Only recently, in the last two decades, have scientists found application of PAI in biomedical imaging, driven by advancements in pulsed laser sources, US detectors, fast data acquisition electronics, and computing power (Das *et al* 2021, Gu *et al* 2023).

The process of PAI can be described by the following steps. (1) Illumination of the tissue with short-pulsed light sources (typically nanosecond lasers) results in local temperature rise (in milli degrees) due to the absorption of light by tissue chromophores. These chromophores can be either intrinsic, such as blood and melanin (Yao *et al* 2010, Chatni *et al* 2012, Wang *et al* 2012b, Zhou *et al* 2012, Lin *et al* 2016, Schwarz *et al* 2016), or extrinsic, such as contrast agents (Pramanik *et al* 2009, Kim *et al* 2010, Ku *et al* 2012, Luke *et al* 2013, Huang *et al* 2016, Brunker *et al* 2017, Guo *et al* 2017, Sivasubramanian *et al* 2017a, Zhang *et al* 2017, Das *et al* 2018, Park *et al* 2018, Zhang *et al* 2018, Miao and Pu 2018, Wang *et al* 2019b, 2019c, Upputuri and Pramanik 2020, Hui *et al* 2022). (2) This local temperature rise leads to a pressure rise through thermoelastic expansion. (3) Propagation of this initial pressure rise as US waves (also known as photoacoustic waves or PA waves) through the tissue. (4) Detection and acquisition of these PA waves at the tissue surface using US detectors. (5) These PA wave data are reconstructed to form images depicting the internal structure/function of the tissue. With these basic steps in mind, one could develop different types of imaging systems by varying light illumination or US detection.

6.2 PAI systems and description

PAI systems consist of two major components—the light source and the US detectors (Xu and Wang 2006). Based on the configuration of the light source with respect to the detector and the imaging resolution of the system, these are majorly classified as tomographic and microscopic systems.

6.2.1 Photoacoustic computed tomography

Photoacoustic tomography (PAT) or photoacoustic computed tomography (PACT) is perhaps the most traditionally thought of system configuration (Upputuri and Pramanik 2017). It has fewer performance limitations when it comes to physically realizing the system. The light source in PAT is often a larger diameter pulsed beam that offers full field illumination. These typically operate at near infrared (NIR) wavelengths where tissue is relatively transparent to enhance the imaging depth (Upputuri and Pramanik 2019). The generated US signals are detected noninvasively by a mechanically scanned single element transducer or an array (linear or curved) of transducers. Using the time-varying signal data along with the speed of sound in different tissues, a planar or three-dimensional image can be reconstructed (Treeby and Cox 2010). The reconstruction technique usually consists of resolving the signals spatially and back projecting to create an image, like a simple delay-and-sum method (Pramanik 2014, Gutta *et al* 2018, Kalva *et al* 2018, Prakash *et al* 2019, Hui *et al* 2023). There are various other types of reconstruction technique, such as Fourier domain reconstruction, system matrix-based reconstruction, iterative method, etc (Treeby and Cox 2010, Schoeder *et al* 2018, Deán-Ben and Razansky 2019, Poudel *et al* 2020, Awasthi *et al* 2021, Yang *et al* 2021). The specifics of the reconstruction depend on the system arrangement and the trade-off between computing resources and precision. In the next section we will briefly introduce two of the most common PAT systems in clinical and pre-clinical use.

6.2.1.1 Linear array photoacoustic tomography

In clinical US imaging, linear array US probes are ubiquitous. Because PAI involves acquiring US signals, sharing a comparable signal acquisition setup is advantageous (Das *et al* 2021). Integrating a light illumination source into an existing linear array-based US imaging system gives a route for simpler translation of PAT for clinical applications. Moreover, the flexibility of linear arrays for free-hand scanning allows for scanning regions that may be inaccessible with other types of transducers. At the time of writing, a variety of companies are manufacturing commercial systems for medical and research PAT imaging (Agrawal *et al* 2019, Singh *et al* 2020, Kratkiewicz *et al* 2021). The only PAI system, Imagio® by Seno Medical, that is food and drug administration (FDA) approved for clinical application is based on linear array transducer.

The challenge of making a dual mode US + PAI based on existing linear array-based US imaging system is that of integrating light delivery and synchronizing the data acquisition and access to raw channel RF data from the US machine to reconstruct the PA images. Unfortunately, not many commercially available clinical US imaging systems have the provision of accessing RF data before beamforming is done. Thus, integrating PAI with those systems is not feasible. However, in the last few years, limited options have been commercially available with access to RF data making them suitable for integration of PAI. One such system with the ability to be retrofitted is the E-CUBE 12R clinical US system (Sivasubramanian *et al* 2017b, 2018) which employs a linear array transducer (L3–12) of 128 elements. A pulsed

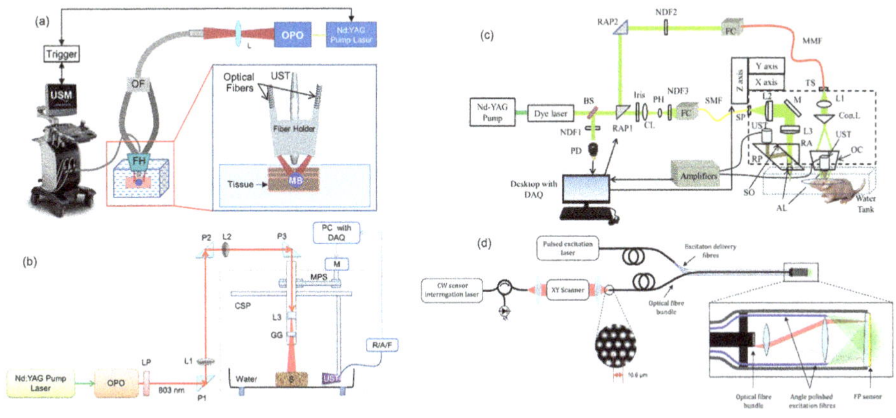

Figure 6.1. PAI systems. (a) Linear array-based photoacoustic computed tomography system. (b) Circular photoacoustic computed tomography system. Adapted with permission from Upputuri and Pramanik (2015). (c) Switchable optical resolution and acoustic resolution photoacoustic microscopy system (Moothanchery and Pramanik 2017) under a Creative Commons License CC BY 4.0. (d) Photoacoustic endoscopy system. Adapted with permission from Ansari *et al* (2020).

light from laser (Nd:YAG pump laser) was focused into a custom-made fiber bundle that has 1600 fibers. The other end of the fiber bundle was bifurcated into 2 bundles of 800 each which ends in a rectangular output, figure 6.1(a). The fiber end and the array transducer were tightly held by a 3D printed probe holder, making it easy for handheld scanning. There could be other methods of light delivery by directly integrating these fiber bundles inside the US transducers during the array transducer manufacturing process.

Linear array-based systems can be easily applied to various clinical applications, but there are some tradeoffs (Shiina *et al* 2018, Yu *et al* 2018, Lin *et al* 2021). It provides 2D imaging view (just like US B-scan imaging) with a restricted field of view compared to full-ring array transducers. On top making 3D volumetric imaging is challenging. To address this, one can employ a translation stage or a mechanical robotic arm with precise position recording to scan the linear array and achieve volumetric imaging (Bunke *et al* 2021, Choi *et al* 2022). Dynamic imaging of a volume is essential for small animal research and clinical translation of PACT. Therefore, for comprehensive 3D volumetric PAI, a different transducer arrangement is preferred.

6.2.1.2 Circular photoacoustic tomography

Circular photoacoustic tomographic systems are characterized by the transducer receiving signals in a circular pattern (Upputuri and Pramanik 2015). Circular PACT is capable of both 2D and 3D imaging. For 2D, the object of interest is placed within the scanning circle and a cross-sectional image of the object is obtained. For 3D imaging, the object is either moved linearly along the axis of rotation or held still while the arrays scan linearly (cylindrical scanning) or rotationally (spherical scanning). 3D tomography is achieved through stacking the planar images along the axis which was scanned. Data acquisition via circular PACT is most commonly done in two ways. Either use a curved or full-ring array transducer or use one or multiple single element ultrasound transducer (UST) rotated around the sample in a circular fashion (Kalva *et al* 2019). Array based systems can have improved imaging speed but are often bulky and more expensive (Yang *et al* 2009b). Single-element systems require a longer time to acquire an image but have the advantage of component simplicity and accessibility.

As shown in figure 6.1(b), the laser beam from an optical parametric oscillator (Continuum, Surelite OPO) pumped by a Nd:YAG laser was passed through a concave lens to expand the beam on the object of interest (Upputuri and Pramanik 2015, Sivasubramanian *et al* 2017c). The UST and the sample were placed in the water tank. The UST was rotated at a 360° by the stepper motor. A data acquisition (DAQ) card was used to acquire the A-lines. A simple delay-and-sum beamforming was used to reconstruct the cross-sectional PA image of the object. Circular transducer arrangements allow for wider coverage for acoustic detection. This gives circular PACT system configurations the capability of detecting PA waves across a broad field of view. As a result, the issue of partial-view detection in the imaging plane is mitigated in comparison with configurations such as linear array-based systems. Due to the geometry of acoustic detection, circular PACT is particularly effective in visualizing cylindrical and oblong targets and for deep tissue imaging, such as the entire body of small animals, and various human extremities.

6.2.2 Photoacoustic microscopy

Photoacoustic microscopy (PAM) can image at a spatial resolution of a few micrometers at a depth of a few hundred micrometers to a few centimeters (Zhang *et al* 2006, Jeon *et al* 2019, Wang *et al* 2021). The advantage of PAM is the combination of optical contrast and ultrasonic/optical resolution. The UST used in PAM is typically a single element focused transducer with a confocal configuration with light illumination. The confocal spot is scanned in 2D (sometimes point-by-point scan using a translational stage or by optical beam scanning using a galvo/MEMS mirror) over the region of interest to acquire the PA A-lines. These A-lines are placed next to each other accordingly to form the 3D volumetric image. Note that usually there is no image reconstruction involved in PAM. Use of maximum amplitude projection (MAP)/maximum intensity projection (MIP) are common to show a 2D projection map of the 3D volume.

Acoustic resolution PAM (AR-PAM) is designed such that the light focus is weak (as AR-PAM image in the diffusive region) and the US transducer focus controls the imaging resolution as the name suggests (Estrada *et al* 2014, Park *et al* 2014, Moothanchery and Pramanik 2017, Moothanchery *et al* 2019, Periyasamy *et al* 2019). This enables deep tissue imaging. Based on the wavelength (532 or 1064 nm) and the center frequency of transducer (50 or 30 MHz) used, the imaging depth ranges from 3 mm to a centimeter (Moothanchery and Pramanik 2017, Periyasamy *et al* 2019). The imaging depth can be further improved by compromising the resolution (use of lower frequency UST like 5–10 MHz). The acoustic focus is achieved by attaching an acoustic lens of suitable radius of curvature to the transducer (or custom-made/off the shelf focused UST is also available) (figure 6.1 (c)) (Moothanchery and Pramanik 2017). The illumination pattern is a doughnut beam (also popularly known as dark-field illumination) which is formed by passing the laser beam through collimating lens, conical lens and the custom-made condenser (Maslov *et al* 2005). The lateral resolution of the AR-PAM operated at 532 nm is ~4.2 and ~130 μm for the set-up optimized for 1064 nm.

Optical resolution PAM (OR-PAM) is designed such that the optical focus is tight, which improves the lateral resolution to ~0.5–5 μm (depending on the optical focal spot size). The imaging depth is limited to ~1–1.5 mm due to the tight focus of the light (beyond certain depth light focusing is challenging in tissue due to light scattering). The scanning head of OR-PAM consists of a beam combiner, which consists of a right-angled prism and a rhomboid prism held together with a thin layer of silicon oil. An acoustic lens is attached to the base of the rhomboid prism. The optical foci and acoustic foci were confocally aligned. Moothanchery *et al* designed a switchable AR-OR-PAM system (figure 6.1(c)). This enabled both the high resolution and low-resolution imaging for a given region of interest.

6.2.3 Photoacoustic endoscopy

Photoacoustic endoscopy (PAE) is designed to bring light illumination and US sensing to the interior of body cavities and hollow organs. Often this is accomplished similar to other endoscopic systems, with a small probe, containing the imaging technology, mounted on an insertion tube or capsule (Yang *et al* 2009a, He *et al* 2019, Guo *et al* 2020, Shabairou *et al* 2020, He *et al* 2023). For example, a PAE system was designed to assist surgical procedures for tumor removal in abdominal cavity (Ansari *et al* 2020). Ansari *et al* fabricated a miniature, all-optical, forward viewing, and minimally invasive PAE (figure 6.1(d)). The outer diameter of the probe was 7.4 mm. A telecentric relay system and a novel light fiber system were integrated into the probe tip. The light was delivered to the tip by a planar Fabry–Perot (FP) sensor through an array of 18 000 optical fibers distributed circumferentially inside the tip. The lateral resolution varied from 60 to 166 μm across the depth of 1–8 mm. Even though PAE has demonstrated promising outcomes in pre-clinical imaging and *ex vivo* imaging of human tissue, *in vivo* human imaging is still a challenge (Lee *et al* 2023). This is due to the strict regulations associated with the imaging devices being inserted into the body cavity. Technical advancements and good manufacturing practices are needed for clinical translation of PAE.

Most of the PAI systems in the research and market can be classified under one of the types of imaging system we described in section 6.2. Next, we will focus on the application of PAI.

6.3 Pre-clinical and clinical applications of PAI

There has been a rapid increase in the number of studies focusing on photoacoustic systems, as evidenced by the growing body of literature (Gu *et al* 2023). The goal and outlook of this literature is often on designing and optimizing imaging systems to achieve higher spatial and temporal resolutions, improved tissue penetration, and reduced artifacts. These advancements in PAI have significantly expanded its applications, ranging from pre-clinical small animal studies to clinical imaging. Pre-clinical applications have grown to comprise of imaging of almost every major organ as well as whole-body imaging on rats and mice. In the pre-clinical PAI field, a significant emphasis is placed on the brain and its vasculatures. This can be attributed to the fact the chromophores in these blood vasculature act as endogenic contrast agents for PA signals. Nevertheless, many other applications are found throughout the body, some of which are highlighted in figure 6.2(a). In the head of a mouse the brain vascular structures, sO_2 map, hemoglobin map, and brain region response due to forepaw simulation have all been imaged and can be seen in figure 6.2(b). Throughout the neck and torso of a mouse the main organs such as common carotid artery (CCA), external jugular vein (EJV), left external jugular vein (LEJV), left internal thoracic vein (LITV), left mammary (LM), left subclavian vein (LSV), left superior epigastric vein (LSEV), right cranial vena cava (RCVC), right internal jugular vein (RIJV), right mammary (RM), right subclavian vein (RSV), right superficial thoracic vein (RSTV), etc, were identified as shown in figure 6.2(c). Cardiovascular imaging of a healthy and myocardial infracted mouse heart were compared in figure 6.2(d). Figure 6.2(e) shows a mammary gland at different wavelengths. In the stomach, nanoprobes have been used for imaging, shown in figure 6.2(f). Healthy versus unhealthy livers were compared and identified as seen in figure 6.2(g). Intestinal ulcers were seen with the use of nanoparticles, which is shown in figure 6.2(h). Lastly the kidney also was analyzed with PACT (figure 6.2(i)). Other animal models that have been investigated include pigs, rabbits, dogs, and

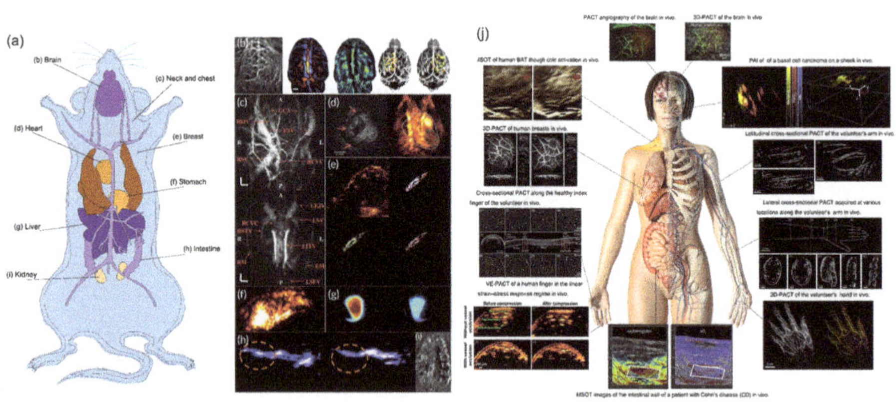

Figure 6.2. Pre-clinical and clinical applications of PAI. (a) Whole-body diagram of a rodent highlighting several regions of application for pre-clinical PAI. (b) Mouse brain vasculature, rat brain hemoglobin and sO_2 concentrations, and functional mouse brain responses to touch stimuli. (c) Rat mid-section PACT images. (d) Mouse heart and mouse heart affected by a myocardial infarction. (e) Mouse mammary gland images at differing wavelengths. (f) Mouse stomach PACT image. (g) MOST liver imaging of healthy liver and steatosis. (h) Mouse intestine ulcer imaging. (i) mouse kidney image. (j) Whole-body diagram of human applications of PAI by body region and their corresponding PAI images. Reprinted with permission from Gu *et al* (2023).

monkeys (Yang and Wang 2008, Hennen *et al* 2015, Lee *et al* 2017, Tian *et al* 2017, Chuangsuwanich *et al* 2018, Liu *et al* 2019a).

Clinical applications are also being researched and progress has been made. The human body contains many regions of interest for PAI including functional neuroimaging, imaging of the ocular structures, teeth, breast, epidermal, vasculature and cardiovascular, extremities and others. A sampling of the uses of PAI on the human body can be seen in figure 6.2(j). Each region will be discussed in detail later. This broader clinical translational potential of PAI surpasses that of purely optical imaging methods, as it has the capacity to deliver comprehensive structural, functional, and molecular information *in vivo* with non-invasive or minimally invasive methods. Depending on the application, different methods and systems of PAI are used.

This section has separated the body into three larger regions: head, trunk, and extremities. Each of these are further separated into subparts according to their organs. The following sections of this chapter will detail the applications of PAI on each organ accordingly.

6.4 Imaging of the head

Starting from the top of the body, the first of the regions that will be discussed is the head. The head is a crucial part of the body and having an imaging modality that can be used for imaging the head becomes a very important tool. However, challenges include thick skull limiting light penetration as well as sound aberration.

6.4.1 Brain imaging

Brain imaging in small animals is one of the major areas of focus and growth for PAI. The strengths of PAI are its high resolution, high optical contrast, and absence of ionizing radiation. These lend themselves well to handling many of the challenges faced by other methods of imaging the brain (Qiu *et al* 2021b). Despite challenges in imaging depth and penetration, PAI allows imaging of both the cortical and deep-tissue vascular networks of *in vivo* rat brains. Furthermore, molecular sensitivity can be achieved using exogenous contrast agents. This can be seen in figure 6.2(b). It shows PAI brain vascular images from Lu *et al* and Zhang *et al* using nanoparticles and dyed droplets as contrast (Lu *et al* 2010, Zhang *et al* 2019). Along with the structural components of the brain such as vasculature and hemodynamics, functional imaging is widely applied in pre-clinical PAI. Cerebral metabolic activity can be quantified by volume of blood flow and changes in oxygen consumption and saturation (Hoge *et al* 1999). These activities cause changes in a tissue's optical parameters, which can in turn be characterized by a PA signal (Grinvald *et al* 1988). Consequently, PAI of the brain can be done using only intrinsic optical absorption properties as an endogenous contrast agent.

Functional imaging of a rat brain was done by Yao *et al* using a PACT system and 2-NBDG and hemoglobin as exogenic and endogenic contrast agents, respectively (Yao *et al* 2013). They were able to capture a brain glucose response to forepaw stimulation. A wearable PAT system was developed by Tang *et al* to capture responses to various sensory stimuli in behaving rats (Tang *et al* 2016). The system comprised of a head mounted probe, with source illumination delivered by a liquid light guide. On the probe was a 3 by 64 array of polyvinylidene fluoride-based acoustic transducer elements. The array was measured at a center frequency of 9.6 MHz and had a bandwidth of 96.4. The light source was a Surlite OPO Plus pulsed laser split into two beams. It had a wavelength range of 680–2550 nm. A custom data acquisition and amplification system received the signals and thus a 3D image was acquired. The responses in the V1 area of the brain were monitored while exposing the rats to a sequence of visual flash stimuli. Averaging 9 trails, the change in HbR in the regions of interest after the onset of a flash was significant. This signals an increase in firing neurons, which consume more oxygen and in turn the HbR in the locality increases while the HbO decreases. Another PACT system was developed for functional imaging on a comparable level with blood oxygenation level dependent (BOLD) functional magnetic resonance imaging (fMRI). Li *et al* were able to show functional changes by measuring systemic sO_2 variations following the inhalation of varying concentrations of oxygen (Li *et al* 2017). Nasiriavanaki *et al* studied the effect of hyerpoxia and hypoxia on the eight main functional regions of the brain with strong correlations in signal region to sensory target region.

Brain tumor identification and characterization is one area where PAI is advantageous. PAI can aid in both early identification and surgical therapy. In surgical removal of a tumor, the surgeon aims to completely remove the malignant tissue. In the midst of irregular and indistinct tumor margins, medical imaging can provide more precise visualizations in comparison to what the human eye can see. Current clinical methods for obtaining these visualizations include MRI guidance techniques and the use of gadolinium, a contrast agent that requires multiple injections due to its short blood half-life. Using different exogenic contrast agents, such as nanoparticles, more comprehensive tumor characterization can be performed *in vivo* (Wang *et al* 2012c, Ni *et al* 2014, Chen *et al* 2016, Gao *et al* 2016, Liu *et al* 2018, McDonald *et al* 2018, Liu *et al* 2019b). Using dye-label contrast agents, Zhang *et al* introduced a dual-scale multi-wavelength PAI system for characterization and analysis of glioblastoma, the most common and lethal cancer in the brain (Li *et al* 2016, Zhang *et al* 2023). They aimed to create an imaging platform where simultaneous spatio-temporal depiction of tumor microvasculature, blood-brain barrier, and immune activity can be done. Their system performed both label-free and with unique contrast agents. The label-free images were able to capture the heterogeneous features of neovascularization in tumor progression while a classic Evans–Blue assay was used to determine dynamic blood-brain barrier permeability and a formulated targeted protein probe (αCD11b-HSA@A1094) for tumor-associated myeloid cells was used to image the contrast of cell infiltration.

In addition to tumors, other diseases such as epilepsy are a subject of interest in pre-clinical PAI applications. The hemodynamic changes associated with epilepsy have been identified using PAI in various ways. Wang *et al* developed a system for real-time 3D imaging of rats (Wang *et al* 2012a). When monitoring the induced seizures according to an acute epilepsy rat model, the focus area was able to be accurately identified (Zhang *et al* 2008). There was an observable association in de-oxygenation and changes in optical absorption at a wavelength of 755 nm in epileptic foci and networks. Brain activity measured by PAI during an epileptic episode could also be correlated with electroencephalography (EEG) spikes (Xiang *et al* 2013, Wang *et al* 2014).

The human brain presents itself as a great challenge to medical imaging. It is a complex organ whose abnormalities and diseases often significantly affect other physiological systems. Considering this importance, there has been a noteworthy amount of attention placed on PAI of the human brain. Small animal imaging applications, some of which are discussed previously, have shown that PAI can both deliver information and augment the information given from other established imaging methods (Qiu *et al* 2021a). In a human head, however, the acoustic aberrations caused by the skull have created considerable challenges (Na and Wang 2021). Different techniques are being proposed to overcome the effects of the human skull, which include characterizing and modeling, deep learning, and computational methods (Huang *et al* 2012). In spite of the skull barrier, PACT brain imaging has been done by Na *et al* (2021). Structural images of the human brain vasculature were taken from volunteers and were compared with 7T magnetic resonance angiography (MRA) images. Figure 6.3(a) shows the PACT imaging set-up. A strong agreement was observed in the brain structures. Functional imaging was then performed using a massively parallel 3D system, which measured deoxyhemoglobin, oxygen, and hemoglobin concentrations. They compared the results of volunteers performing sensory tasks to fMRI images. A sample of the imaging results can be seen in figure 6.3(b). The temporal activation in the region of interest was recorded and compared to the BOLD signal. There was once again a strong agreement, providing evidence for the feasibility of high-speed 3D PACT for functional brain imaging.

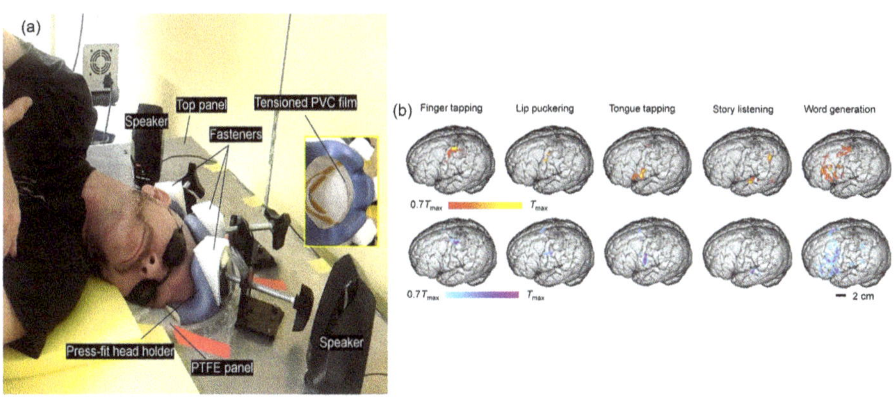

Figure 6.3. 3D fPACT human brain imaging. (a) PACT imaging system positioning and set-up. (b) Active regions indicated by fMRI (top) and fPACT images (bottom) from various sensory stimuli. Adapted with permission from Na *et al* (2021) © 2024 Springer Nature Limited.

6.4.2 Cheek epidermal imaging

As the organ on the surface of the body, the human epidermis provides ease of access for various PAI applications. Skin-related diseases are found throughout the body and will be discussed in more detail in later sections. Specifically on the human face, PAI has been applied to skin cancer characterization and identification. Chuah *et al* created a handheld clinical system for 3D volumetric imaging of basal cell carcinoma (BCC) (Chuah *et al* 2019). Using volumetric multispectral optoacoustic tomography (vMSOT), label-free images of the cheeks and forehead were acquired. The tissue boundaries were able to be clearly identified according to the distribution of melanin and the oxy-hemoglobin signals. From this the tumor dimensions could be found.

6.4.3 Teeth inspection and health

In teeth, discoloration often means tissue decay or potential onset of decay. Using PAI to detect this has been a point of interest in medical imaging as PAI plays well with detection of color changes in tissue. Human teeth have been imaged *ex vivo* to show that PAI is a feasible method for diagnosis of dental caries, cracks, and lesions (Hughes *et al* 2015, Cheng *et al* 2016, Windra Sari *et al* 2023). In cases where the caries is hidden by the multifaceted surface inherent to teeth, PAI technology can be especially helpful (Tasmara *et al* 2023). Beyond caries, PAI has also been used to determine tooth decay (El-Sharkawy and El Sherif 2012). In an experiment, teeth in a state of decay were imaged with a Nd:YAG laser system and compared to the images of healthy teeth. There were substantial differences between the healthy and damaged tissues.

For *in vivo* PAI imaging of teeth, different transducer designs are being researched to better adapt to the mouth area. These designs depend on the application and the source configuration. Fu *et al* achieved imaging of the periodontal tissue of the posterior teeth using a toothbrush-like probe design and a laser diode (Fu *et al* 2022). Fu *et al* then created a hockey-stick transducer design with a 680–790 nm source. This allowed imaging of teeth farther back in the jaw. All but the last two molars were able to be reached. The target was to image periodontal pockets, which were found to produce higher intensity signals. Further investigation of gingivitis using a contrast agent in a pig mouth was also done with notable results. Gingivitis infection has been identified using a dual mode fluorescent and photoacoustic system. Moore *et al* performed imaging of gingipain proteases from *P. gingivalis,* which are an infection biomarker. A contrast agent was designed to fluoresce in interactions with the gingipain proteases. This fluorescence can be identified by PAI, accurately identifying infection in human samples as well as gum sites and pocket depths in swine (Moore *et al* 2022).

6.4.4 Eye disease identification

In a clinical setting PAI has potential to provide valuable information for monitoring and treatment of various ocular diseases. However, eyes and the photoreceptors contained within are innately sensitive to light. This along with the continuous motion, and optical effects of the cornea and crystalline lens of the eye, make them a challenging organ to image (Korenbrot 2012, Liu and Zhang 2016). Simultaneous imaging of the various ocular components has yet to be achieved in the human eye. Using excised goat tissue, Samanta *et al* were able to show that PAI has the capability to capture the entire ocular structure while clearly capturing the major structures (Samanta *et al* 2023). These include the cornea, aqueous humor, iris, pupil, eye lens, vitreous humor, and retina. Deeper structures were viewed in a mouse eye in a study from Haindl *et al.* Rose Bengal was used as a contrast agent and imaged by a multimodal ophthalmoscope (Haindl *et al* 2023). This enabled visualization of the choroidal vasculature and sclera in mice for the first time.

6.4.5 Thyroid imaging

Tumors of the thyroid are common diseases found in the head and neck. Thyroid tumors are often benign or don't require immediate treatment. Thus, characterization of the relevant tissue area is important to prevent both underdiagnosis and overdiagnosis (Vaccarella *et al* 2016). A non-invasive functional imaging modality such as PAI is beneficial for early diagnosis, management, and treatment. Custom systems have been developed to identify important features in the thyroid and any cancerous tissue that may be present. Dima *et al* used a curved array system to map the thyroid outline and vasculature (Dima and Ntziachristos 2016). In comparison to colored Doppler US, PAI was more efficient with a similar high-quality image. Yang *et al* performed thyroid imaging with a linear array transducer set-up (Yang *et al* 2017). They used a modified clinical US system to find structural information with precision. Adding more important information for diagnosis, Kim *et al* used multispectral acquisition to perform imaging and analysis on *in vivo* thyroids. From the data the oxygen saturation of thyroid nodules was calculated. Benign and papillary thyroid cancer cases were differentiated according to a multiparametric statistical method. The probability of correct classification with respect to combined PAI data and the American Thyroid Association Guideline achieved a sensitivity of 83% and a specificity of 93%.

6.5 Applications of the trunk

Next, we will move on to the middle section of the body and discuss PAI applications of the trunk. The trunk is the thickest part of the body, limiting the imaging depth; presence of bones are some of the challenges one encounters while doing PAI of the trunk area.

6.5.1 Cardiac imaging

The heart, a crucial organ in the human body, plays an essential role in circulating blood, supplying oxygen and nutrients to the body's tissues. The vitality of the human heart cannot be overstated due to its pivotal role in the overall operation of the body. Over an average lifespan, the heart pulsates approximately 2.5 billion times, propelling millions of gallons of blood to every corner of the body.

This constant circulation delivers oxygen, nutrients, hormones, various compounds, and a multitude of crucial cells. It also expels the byproducts of metabolism. In the realm of cardiology, *atherosclerosis* is a major cardiovascular disease (CVD) (Lnsis 2000, Wu *et al* 2021). Atherosclerosis is a progressive disease marked by the accumulation of lipids and fibrous elements in large arteries. It begins with the formation of foam cells, cholesterol-filled macrophages, beneath the endothelium. These fatty streaks appear first in the aorta, then in the coronary arteries, and later in the cerebral arteries. Although not clinically significant, they are precursors to more advanced lesions known as *fibrous lesions*. These lesions consist of a *fibrous cap* of smooth muscle cells and extracellular matrix encapsulating a lipid-rich *necrotic core*. Over time, these *plaques* can develop calcification, surface ulceration, and hemorrhage from small vessels growing into the lesion from the vessel wall's media. If untreated, these plaques can form blood clots, leading to myocardial infarction or stroke (Lnsis 2000).

The detection of vulnerable plaques is essential for guiding cardiovascular interventions and preventing acute cardiac events. Clinics routinely use a variety of imaging techniques for CVD diagnosis, including US, MRI, CT, PET, and SPECT (Shishikura 2016). Each of these techniques has its limitations: CT imaging involves high radiation doses; MRI is relatively expensive and not always readily available. In contrast, US imaging is safe, user-friendly, and known for its high resolution, affordability, and accessibility, making it the most common diagnostic imaging technique in cardiology. In recent years, there have been significant advancements in PAI for diagnosing plaques. The vulnerability of plaques is closely related to their composition, typically including lipid, calcification, intraplaque hemorrhage, and macrophages (Gao *et al* 2007, Naghavi and Falk 2010). These components can be effectively visualized by PAI, making it a powerful tool for characterizing vulnerable plaques. PAI has become a significant choice for plaques intervention in recent years, with two main approaches: endoscopic catheter-based PAI (also known as *intravascular PA* or *IVPA* imaging) (Jansen *et al* 2014, Wu *et al* 2014, Wu *et al* 2016) and non-invasive PAI (Arabul *et al* 2017). Moreover, the multispectral PAI, which operates at different optical spectral ranges, can measure the oxygen saturation in the heart tissue, providing vital information about the heart's oxygen supply and demand. PAI is intrinsically linked to US imaging, making it a promising new imaging technique for clinical applications in cardiology. As graphical processing unit (GPU) technology continues to evolve, deep learning (DL) methodologies can be successfully combined with PAI to improve both image clarity and understanding. This approach has already demonstrated significant improvements in US imaging within the field of cardiology (Gao *et al* 2017, Madani *et al* 2018).

6.5.1.1 IVPA imaging of vulnerable human atherosclerotic plaques

Emerging as a promising tool in the field of CVD, intravascular photoacoustic (IVPA) imaging has demonstrated its potential in detecting and characterizing vulnerable plaques (Jansen *et al* 2014, Wu *et al* 2014, Wu *et al* 2016). A distinguishing characteristic of these plaques is a lipid-rich core. When lipids (an endogenous contrast agent), which absorb light at specific wavelengths, are exposed to pulsed light, they emit a PA signal. This signal is captured and utilized to form an image of the plaque. The capacity of IVPA imaging to detect lipids is of particular significance as it offers a direct assessment of the plaque's lipid content—a key determinant of its vulnerability. This accurate identification of the lipid core's presence and size, enables clinicians to evaluate the risk of plaque rupture and determine the most suitable treatment approach. Furthermore, IVPA imaging can supply this crucial information in real time during a catheterization procedure, offering immediate insights to the clinician. This could potentially enhance the effectiveness of interventions and lead to improved patient outcomes.

In 2011, Jansen *et al* (2011), reported the first IVPA imaging of human atherosclerotic coronary arteries *ex vivo*. They made a hybrid IVPA/IVUS catheter, which could co-register PA and US at the same time as shown in figure 6.4(a). The system was composed of a core optical fiber with a diameter of 400 μm and a lead–zirconium–titanate (PZT) US transducer with a diameter of 1.0 mm. The working tip of the catheter, comparable in size to the edge of a 10 Eurocent coin, is depicted in figure 6.4(b). IVPA/IVUS was performed on an advanced lesion (left anterior descending artery, 56-year-old male), and the histological data in figure 6.4(c) reveals circumferential intimal thickening, a large eccentric lipid-rich lesion, a calcified area,

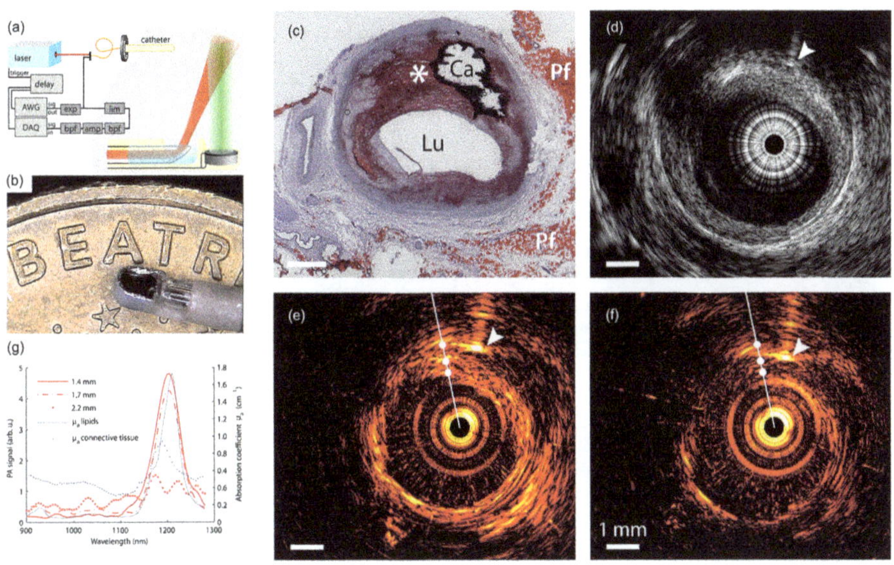

Figure 6.4. IVPA imaging. (a) Diagram of the IVPA setup, including a detailed schematic of the catheter tip, showing the beam layout. AWG, arbitrary wave generator; DAQ, data acquisition; exp, expander; lim, limiter; bpf, bandpass filter; amp, amplifier. (b) Photograph of the working catheter tip of IVPA; Co-registered IVPA/IVUS imaging of an advanced human atherosclerotic plaque. (c) Histology: Oil Red O stain shows the presence of a lipid-rich plaque (*) as well as a calcified area (Ca). Lu, lumen; Pf, peri-adventitial fat. (d) IVUS image; IVPA images obtained at (e) 1210 nm (high lipid absorption) and (f) 1230 nm (low lipid absorption). Arrowheads indicate the needle used for marking. (g) PA spectra at three locations on the white line shown in panels (e) and (f). Adapted with permission from Jansen *et al* (2011) © Optica Publishing Group.

and regions of peri-adventitial fat. This morphology was confirmed by the IVUS imaging in figure 6.4(d). The IVPA image (figure 6.4(e)) at 1210 nm displayed a bright signal from the intimal border and deeper tissue layers in the eccentric plaque and the peri-adventitial fat in the bottom right corner. However, at 1230 nm (figure 6.4(f)), the signal in these areas was significantly reduced, aligning with the absorption spectrum of lipids in this wavelength range. PA spectra were collected along image lines that sampled the plaque tissue. Figure 6.4(g) presents spectra at three locations. The sites at distances of 1.4 and 1.7 mm from the catheter were situated within the lipid-rich plaque, and their corresponding PA spectra closely matched the lipid absorption reference. The site at 2.2 mm is just outside the plaque area.

6.5.1.2 Non-invasive PA imaging for cardiovascular diagnosis

Given its sensitivity to various forms of hemoglobin, PAI serves as a non-invasive and cost-effective technique for identifying vulnerable plaques with intraplaque hemorrhages. Additionally, it can measure extra cardiovascular hemodynamics, such as blood flow and oxygen saturation, thereby aiding in the precise diagnosis and prevention of CVDs. Recently, in 2021, Muller *et al* (2021) demonstrated the first *in vivo* intra-operative PAI of intraplaque hemorrhages in human carotid artery. They made a fully integrated handheld laser-diode based PA/US probe as shown in figure 6.5(a), to demonstrate the capability of PAI to image plaques with intraplaque hemorrhages (which are key indicators for vulnerable plaque) in human, by performing the first pilot clinical study of PAI in patients during carotid endarterectomy (CEA) surgery. From figures 6.5(b–e) that a strong PA response related to

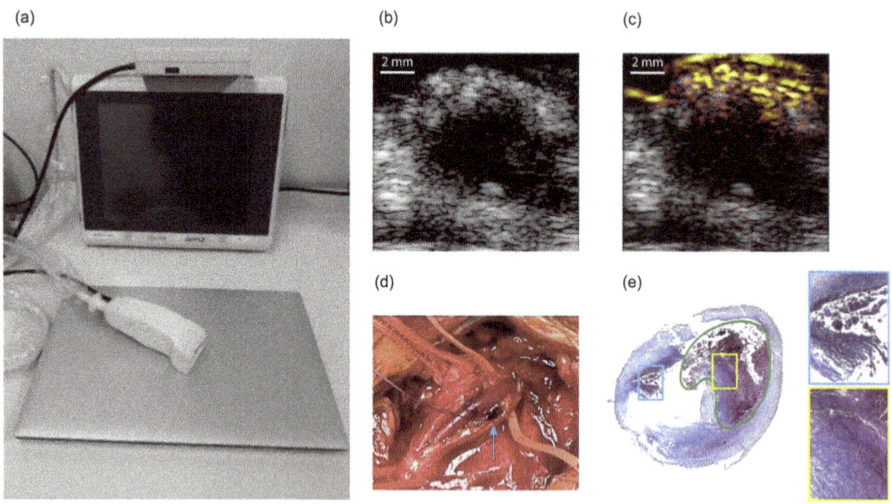

Figure 6.5. Carotid artery imaging (a) The portable PA/US imaging system, consisting of a PA/US scanner, a laptop and the handheld PA/US probe, is shown. *In vivo* PA and US image of a human carotid artery with intraplaque hemorrhage. (b) US image. (c) Combined PA/US image (808 nm). (d) Photo of the carotid plaque during CEA surgery. (e) Masson's trichrome staining of the artery. The region marked in green is a lipid core filled with a significant hemorrhage. The highlighted boxes point out two regions of hemorrhages found in the plaque. Adapted with permission from Muller *et al* (2021) © Optica Publishing Group.

the presence of the intraplaque hemorrhages (figure 6.5), and a diffused signal pattern was observed in the hemorrhage lesion, probably caused by the heterogeneity in the composition of the plaque (Muller *et al* 2021).

6.5.2 Breast cancer diagnosis

Breast cancer has become the most frequently diagnosed cancer worldwide (Manohar and Dantuma 2019), making up 1 in 8 of all cancer diagnoses and totaling 2.3 million new cases across both genders. It constituted a quarter of all female cancer cases and was the most diagnosed cancer among women in 2020 (Arnold *et al* 2022). Its impact has been escalating globally, especially in transitioning countries. In 2020, an estimated 685 000 women succumbed to breast cancer, accounting for 16% or one-sixth of all female cancer deaths. The World Health Organization (WHO) recently initiated the Global Breast Cancer Initiative in response to the previously inadequate public health measures. The WHO, along with global partners, is striving to decrease breast cancer mortality by promoting early diagnosis, appropriate treatment, and effective patient management.

Breast cancer physically originates in the milk ducts or the milk-producing lobules of the breast, manifesting symptoms such as a new lump in the breast or underarm, nipple discharge or itching, and changes in the texture of the breast or nipple. From a psychological perspective, the diagnosis and treatment of breast cancer can induce considerable stress and anxiety. Socioeconomically, it can impose a significant financial strain due to the high costs of treatment and loss of productivity. The etiology of breast cancer is complex, encompassing factors like being female, advancing age, obesity, alcohol consumption, a family history of breast cancer, exposure to radiation, reproductive history, tobacco use, and postmenopausal hormone therapy. Interestingly, about half of all breast cancer cases occur in women who do not have any specific risk factors apart from their sex and age. Imaging is a crucial component in the comprehensive management of breast cancer, serving key functions in detection, diagnosis, monitoring of neoadjuvant therapy, directing biopsies and surgeries, and post-operative observation (Lashof *et al* 2001, Kopans 2007, Benson *et al* 2009).

In the realm of breast cancer detection, x-ray mammography is primarily used for screening, while diagnosis involves clinical examination, x-ray imaging, US imaging, and image-guided needle biopsy. MRI is employed when findings from x-ray and US imaging are uncertain (Kuhl 2007). However, these methods have limitations. X-ray and US imaging lack optimal sensitivity and specificity (Britton *et al* 2009, Vinnicombe *et al* 2009, Hooley *et al* 2011). X-ray mammography involves ionizing radiation, painful breast compression, and performs poorly in radio-dense breasts (Pediconi *et al* 2009, Brédart *et al* 2012). US imaging is limited by high false positive rates and operator variability (Hooley *et al* 2011). MRI, despite its high sensitivity, good spatial resolution, and lack of ionizing radiation, suffers from limited specificity, requires contrast agents, and is logistically challenging due to the need to time imaging with the menstrual cycle in pre-menopausal women (Kuhl 2007, Onesti *et al* 2008). It's also expensive, not universally available, and certain patients are excluded due to factors like claustrophobia and pacemakers (Essink–Bot *et al* 2006). Angiogenesis, the formation of new blood vessels, is a key characteristic of cancer, occurring early to support the growth of invasive cancers (Hanahan and Weinberg 2011). This process leads to an increase in local microvascular density, resulting in abnormal, dilated, and tortuous vessels (Carmeliet and Jain 2011). These newly formed vessels contain hemoglobin (Hb) and its oxygenated variant (HbO_2), both of which have strong and specific optical absorption spectra. This presence of Hb and HbO_2 in the enhanced vascularization provides a distinct optical absorption contrast between cancerous and healthy tissue. The PAI technique, which relies on optical absorption in tissue, appears to be an optimal choice for breast cancer imaging. With its ability to use multiple wavelength-based excitation, PAI is effective in detecting such angiogenesis-driven optical absorption contrast of tumors, making it a powerful tool in the fight against breast cancer.

The concept of using NIR-PA for breast imaging was initially proposed in 1994 by Oraevsky *et al* (1994). However, it took 7 years for them to first demonstrate its application in 2001 (Oraevsky *et al* 2001). Since then, numerous prototypes for breast imaging have been developed and reported (Manohar and Dantuma 2019). Based on the geometry of the imager, the PAI systems for breast cancer diagnosis can be broadly classified into two types: (a) *local breast imaging based on intensity projection* (Heijblom *et al* 2015), and (b) *whole-breast imaging* (Manohar *et al* 2005, Lin *et al* 2018).

In the year 2015, a study was conducted by Heijblom *et al* (2015), where they used a Twente Photoacoustic Mammoscope (PAM) to perform PAI on a human breast affected by infiltrating ductal carcinoma (IDC). IDC is a form of cancer that originates in the epithelial cells lining the lobules and ducts of the breast. This is depicted in figure 6.6(a). Furthermore, they compared the results obtained from PAM with those from magnetic resonance (MR) images. They observed a strong correlation between the lesion locations in both the PAM and MR images.

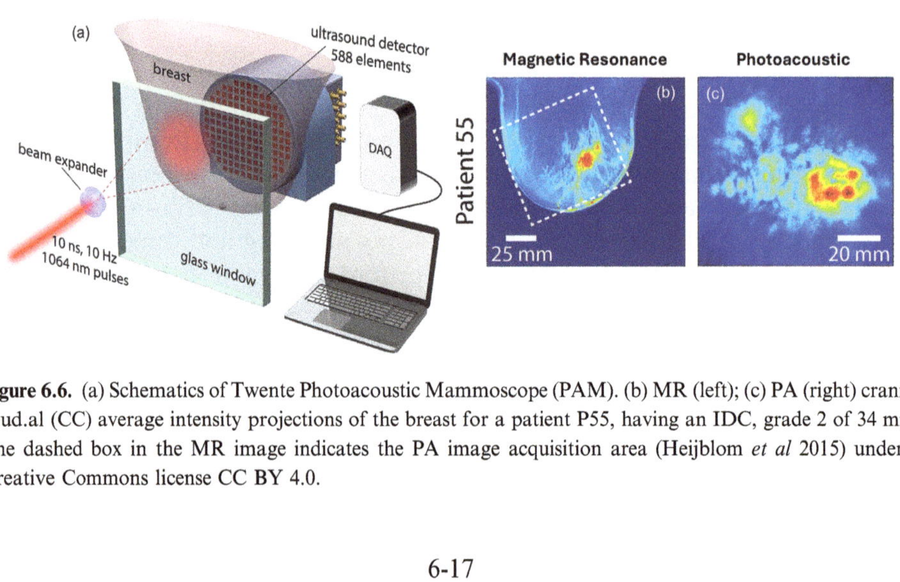

Figure 6.6. (a) Schematics of Twente Photoacoustic Mammoscope (PAM). (b) MR (left); (c) PA (right) cranio-caud.al (CC) average intensity projections of the breast for a patient P55, having an IDC, grade 2 of 34 mm. The dashed box in the MR image indicates the PA image acquisition area (Heijblom *et al* 2015) under a Creative Commons license CC BY 4.0.

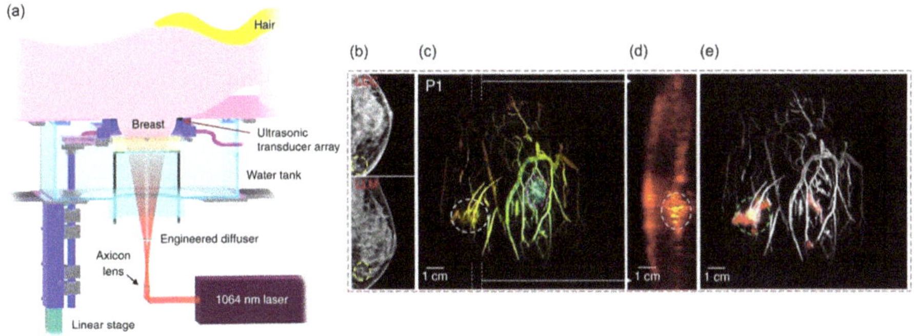

Figure 6.7. Schematic of the SBH-PACT system. (a) Perspective cut-away view of the system with patient on bed; SBH-PACT of cancerous breasts. (b) X-ray mammogram of the affected breasts of breast cancer patient (P1–48-year-old female patient with an invasive lobular carcinoma (grade 1/3)), (LCC left cranial-caudal, LLM left lateral-medio). (c) Depth-encoded angiograms of the affected breast acquired by SBH-PACT. Breast tumors are identified by white circles. (d) MAP images of thick slices in sagittal planes marked by white dashed lines in (c). (e) Automatic tumor detection on vessel density maps. Tumors are identified by green circles. Background images in gray scale are the MAP of vessels deeper than the nipple. (Lin *et al* 2018) under a Creative Commons license CC BY 4.0.

Interestingly, they also found a striking similarity in the overall appearance and shape of the lesion, as can be seen in figure 6.6(b–c).

However, intensity projection-based systems can only image a limited portion of the breast and may produce artifacts due to the limited view. As a result, there's a growing interest in PA whole-breast imaging, which can provide a more comprehensive view of the breast (Lin *et al* 2018). In 2018, Lin *et al* developed a system known as single-breath-hold photoacoustic computed tomography (SBH-PACT) that can unveil intricate angiographic structures within human breasts as shown in figure 6.7(a–e) (Lin *et al* 2018).

The SBH-PACT system boasts a substantial penetration depth of 4 cm *in vivo*, coupled with high spatial and temporal resolutions (255 μm in-plane resolution and a 2D frame rate of 10 Hz). The entire breast can be scanned within a single breath hold of approximately 15 s, allowing for the capture and subsequent 3D back-projection reconstruction of a volumetric image with minimal motion artifacts induced by breathing. By identifying areas of increased blood vessel density associated with tumors, SBH-PACT can effectively reveal tumors, demonstrating its potential for high sensitivity in breasts that are radiographically dense.

6.5.3 Photoacoustic imaging in gastrointestinal (GI) diagnosis

Gastrointestinal (GI) diseases cover a broad spectrum of conditions impacting the GI tract, which extends from the mouth to the anus. In comparison to 1990, the worldwide figures for incident cases, fatalities, and Disability-Adjusted Life Years (DALYs) related to GI diseases saw a significant rise in 2019. Specifically, there was a 74.44% increase in incident cases, a 37.85% rise in deaths, and a 23.46% surge in

DALYs (Wang *et al* 2023). These include various cancers (Ono *et al* 2021) that can develop in any part of the GI tract, such as the esophagus, stomach, liver, pancreas, and colon. These cancers are often associated with lifestyle factors like diet, smoking, and alcohol consumption, and their symptoms might not manifest until the cancer has progressed significantly, complicating early detection. Conditions like inflammatory bowel disease (IBD) fall under the category of inflammatory diseases, leading to chronic inflammation in the digestive tract (Seyedian *et al* 2019). IBD typically emerges in adolescence or early adulthood, but it can also appear later in life. Symptoms can range from diarrhea, abdominal cramps, and fatigue, bloating, bloody stools, loss of appetite, or weight loss. Various diseases can cause lesions in the GI tract. For instance, skip lesions, a characteristic of Crohn's disease, occur when parts of the intestine are inflamed, interspersed with patches of normal tissue. Another type of lesion, known as a cobblestone ulcer, is a small, densely packed lesion that can develop in the GI tract of individuals with Crohn's disease (Parray *et al* 2011). These lesions can lead to symptoms such as abdominal pain, nausea, vomiting, and diarrhea.

Early diagnosis of such GI diseases is crucial as these conditions can be extremely uncomfortable and, in some cases, even life threatening. If left untreated, GI issues can lead to dehydration, malnutrition, and even serious infections. Currently, the diagnosis of GI diseases involves various methods such as endoscopy (Teh *et al* 2020), colonoscopy (Gangwani *et al* 2023), biopsy (Han *et al* 2014), CT scan (Macri *et al* 2023), x-rays (Lee *et al* 2010), and MRI (Maccioni *et al* 2023). Among these expensive, harmful radiation-based, invasive techniques, white light endoscopy (Mannath and Ragunath 2016) stands out as a minimally invasive technology that is extensively utilized in the field of clinical gastroenterology. To enhance the visualization of the GI track, PAE imaging has recently been introduced (Guo *et al* 2020, Ikematsu *et al* 2022, Liang *et al* 2022, He *et al* 2023). This technology improves contrast and depth of penetration, enabling the visualization of features previously unseen in endoscopic procedures.

So far, the capabilities of various PAE probes have been tested on *in vivo* animals or excised tissues (He *et al* 2018, Liang *et al* 2022). Recently, in 2022, Liang *et al* (2022) employed PAE imaging probe (figure 6.8(a)) to *in vivo* image a rat rectum and visualize the oxygen saturation changes during acute inflammation, which is difficult with other imaging modalities. Each B-scan contained 2000 A-lines, with an angle step of 0.1 degrees, and an imaging speed of 1 Hz. The laser pulse energy was maintained at 250 nJ. The PAE imaging in figure 6.8(b) reveals a high-resolution, three-dimensional distribution of blood vessels, along with the oxygen saturation levels in the rectum. The imaging outcomes display markedly distinct profiles of vascular networks. Region #1 presents numerous vessels running parallel in the longitudinal direction. Region #2 displays a relatively uniform, grid-like network of vessels. Region #3 demonstrates branch-like structures, comprising several main vessels encircled by numerous branch vessels and capillaries (Liang *et al* 2022). While the research validates the potential of PA endoscopy in small animals, recent advancements are focusing on developing capsule-based PAE systems suitable for use in humans (He *et al* 2019, Ali *et al* 2022).

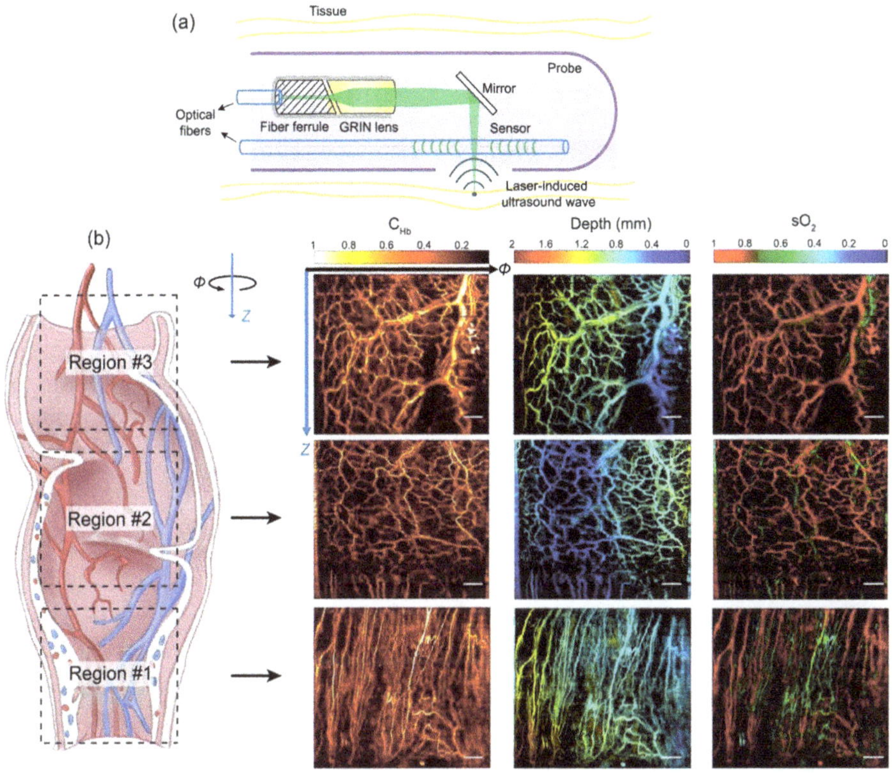

Figure 6.8. (a) Schematic of the PAE probe. (b) *In vivo* endoscopy imaging results of the hemoglobin concentration (CHb), depth, and oxygen saturation (sO_2) of a rat rectum in three different regions. (Liang *et al* 2022) under a Creative Commons license CC BY 4.0.

6.5.4 Gynecological monitoring

Among the various gynecological disorders and diseases, *ovarian cancer* is recognized as the most lethal of all gynecological malignancies (Nie *et al* 2023). This is attributed to several reasons, including its tendency to remain undiagnosed until it has metastasized within the pelvis and abdomen. At this advanced stage, ovarian cancer becomes more challenging to manage and is frequently fatal. Ovarian cancer in its early stages, where the disease is restricted to the ovary, has a higher likelihood of successful treatment. However, early-stage ovarian cancer seldom exhibits any symptoms. This underscores the importance of regular medical diagnosis. Over the past 30 years, the global burden of ovarian cancer disease has seen a significant increase. In 2019, there were 294 422 incident cases, 198 412 deaths, and 5.36 million DALYs. Compared to 1990, this represents an increase of 107.8% in new cases, 103.8% in deaths, and 96.1% in DALYs (Zhang *et al* 2022). Despite these increases, the age-standardized rates remained stable from 1990 to 2019.

Present clinical diagnostic techniques, such as transvaginal ultrasonography (TUS), Doppler US, and so on, have been found to lack the sensitivity and specificity needed for early detection of ovarian cancer (Kobayashi *et al* 2008,

Temkin *et al* 2017, Andreotti *et al* 2020). Furthermore, a significant number of benign ovarian lesions may be challenging to accurately diagnose using current imaging technologies, resulting in increased healthcare costs and unnecessary surgeries. In order to address these challenges, researchers are turning their attention to the non-invasive PAI technology, which is showing promising results in identifying functional biomarkers associated with tumor angiogenesis and the consumption of blood oxygenation, or tumor metabolism (Nandy *et al* 2018, Amidi *et al* 2021, Nie *et al* 2023).

In 2023, Nie *et al* developed a coregistered US and PAI protocol for the transvaginal imaging of human ovarian lesions by combining a clinical US transvaginal probe, a custom-made optical fibers holder for light delivery, and a tunable laser, as shown in figure 6.9(a) (Nie *et al* 2023). They employed this tool to diagnose a 46-year-old woman with bilateral cystic lesions. Figure 6.9(b) shows the US image of the right ovary, with a simple cyst of 4.2 cm in diameter. Figure 6.9(c) illustrates the PAT total hemoglobin concentration (rHbT) map overlaid on the coregistered

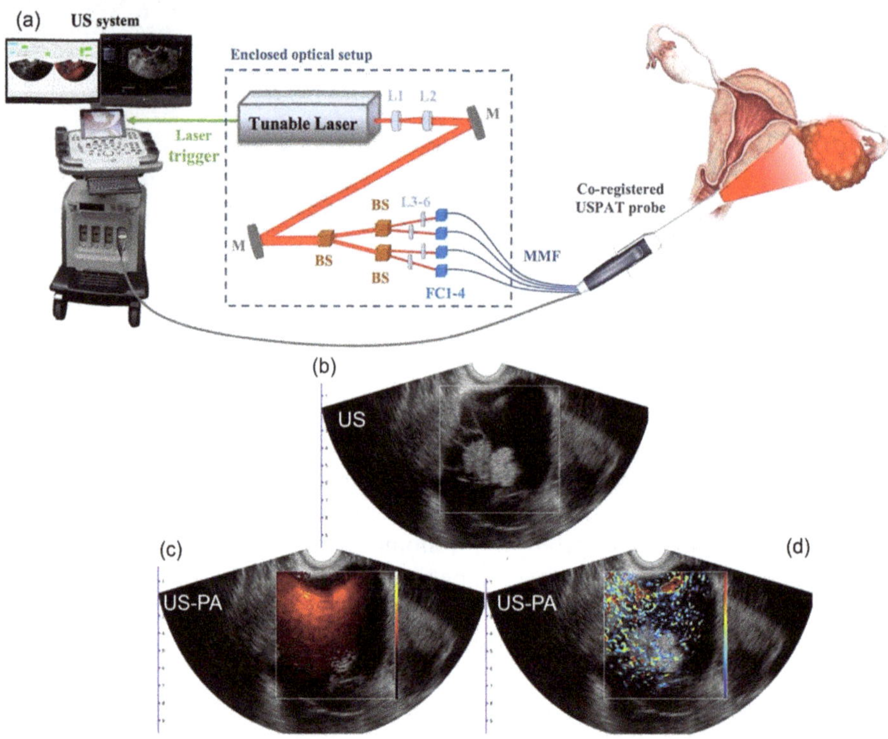

Figure 6.9. (a) Schematic design of the coregistered US and PAT system and probe. (Plano-convex lens (L1), plano-concave lens (L2), beam splitters (BS), and multi-mode fibers (MMF), four plano-convex lenses (L3–6) and fiber couplers (FC1–4), and Mirrors (M)). (b) US of the right ovary of a 46 year old woman with bilateral cystic lesions. (c) The PAT rHbT map superimposed onto the coregistered US showing scattering signals on the left side of the lesion with a low average level of 4.8 (a.u). (d) The %sO_2 overlaid on US image. The depth was marked on the right side of the B-scan images. Adapted with permission from Nie *et al* (2023) © 2024 MyJoVE Corporation.

US, revealing scattering signals on the left side of the lesion with a low average level of 4.8 (a.u.). Whereas figure 6.9(d) depicts the blood oxygen saturation (sO_2) map overlaid on the coregistered US image, which indicates a higher sO_2 content of 67.5%. The surgical pathology report confirmed a normal right ovary with follicular cysts (Nie *et al* 2023).

6.6 Photoacoustic imaging of the extremities

Moving to the bottom and outside of the body we will discuss the applications of PAI on the extremities.

6.6.1 Finger imaging

The human finger is widely imaged using photoacoustics for better understanding of rheumatoid arthritis and study the blood vessels (Song *et al* 2010, Favazza *et al* 2011, Nishiyama *et al* 2019). Linear array-based systems, curvilinear geometry and single array-based transducers were used for imaging of the human finger (Xu *et al* 2013, Jo *et al* 2017, van den Berg *et al* 2017). Nishiyama *et al* designed a ring array that could image objects of diameter 0.1–0.5 mm (Nishiyama *et al* 2019). The finger was placed in the ring array sensor, which has the optical illuminator. 800 nm light generated by the optical parametric oscillator pumped by the Nd:YAG laser was used for the illumination. Average light intensity on the illuminator surface was ~16.5 mJ, which translates to the optical fluence of ~2.3 mJ cm^{-2}. The 256-element ring array was of diameter 33 mm. 0.4, 0.3 and 5 mm was the pitch, height, and width of each of the elements, respectively. Center frequency was 2.8 MHz and the fractional bandwidth was 85%. The scanning of the human finger was done over 30 mm at a scanning interval of 0.1 mm. The PA-US cross-section of the finger along the *xz* and *yz* views were studied. The vasculature of the finger demonstrated here was ~0.26 mm in diameter and ~4 mm deep from the skin surface. Distal interphalangeal and proximal interphalangeal joints of a healthy subject were studied because these are the joints which have inflammation during rheumatoid arthritis. Some artifacts were observed due to the alignment of the finger in the imaging system and a few were due to the use of uniform speed of sound during reconstruction.

Ahn *et al* developed a high-speed, high-resolution, spectroscopic PAM for structural and functional imaging of micro-vessels in the human finger (Ahn *et al* 2021). This system was used to study the spatial displacement of the blood vessels during arterial pulsation. Initially a 2D scan was done to localize the blood vessels, then these blood vessels were monitored for the oxygen consumption during brachial cuffing. Blood perfusion was also observed after the release of the cuff. 532 and 559 nm nano seconds pulsed lasers were used for illumination (figure 6.10(a)). This dual-wavelength imaging is crucial for spectral unmixing. The laser beam was delivered to the optical resolution photoacoustic microscopy (OR-PAM) system by single-mode fiber. Water immersible micro-electro-mechanical system (MEMS) based opto-ultrasound combiner and scanning mirror was used to focus the laser on the target. A 50 MHz US transducer was used to acquire the PA signals. 5 and 30 μm was the

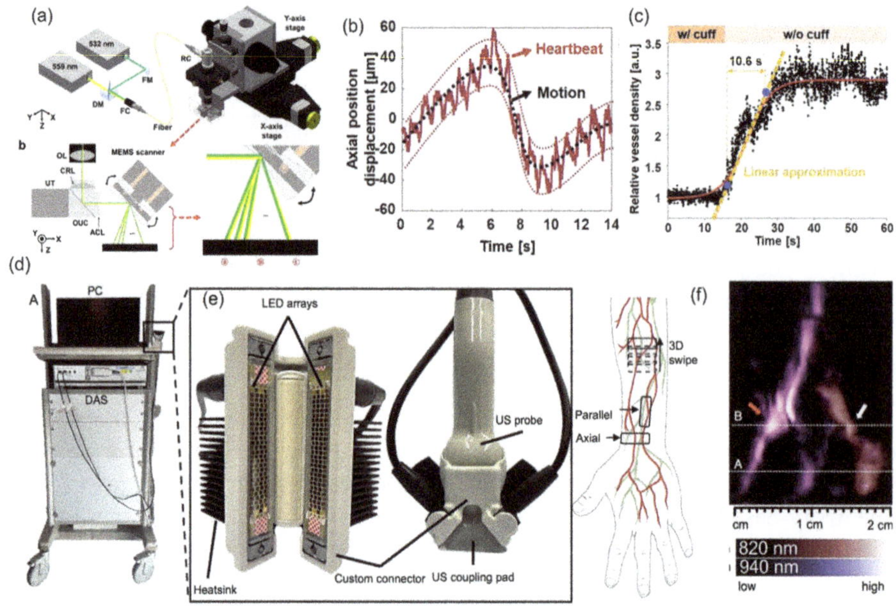

Figure 6.10. (a) Dual wavelength OR-PAM system with the scanning head. (b) Axial displacement of the blood vessel with respect to time. (c) Change in blood vessel density with respect to time. Adapted with permission from Ahn *et al* (2021). (d) Handheld photoacoustic tomography system. (e) Imaging orientation on the forearm. (f) Blood vessel and lymph vessel in the forearm. Adapted with permission from Van Heumen *et al* (2023).

lateral and axial resolution of the system, respectively. The imaging speed was 50 Hz which was sufficient to monitor the heart rate. Signal-to-noise (SNR) calculated from *in vivo* finger image was 28 dB. Optical fluence on the skin was 14 and 16.6 mJ cm^{-2} for 532 and 559 nm, respectively. Firstly, the arterial pulsation was monitored. For this step, the vascular movement was quantified by selecting the region of interest, which is a single blood vessel, then the pixel with highest PA signal was detected. The displacement of the highest intensity pixels over the consecutive images was representative of the heartbeat and motion (figure 6.10(b)). The imaging of this movement was feasible. The heartbeat computed from the PA images matched well with the heartbeat recorded by the smartphone sensor.

Next, to study the dynamic changes, the blood vessels were monitored during the release of the brachial cuff. On monitoring the MAP images, more blood vessels were observed once the cuff was released. The blood vessel density was quantified on the skin before and after realizing the cuff. On deflating the cuff, the blood vessels were 2.8 times denser (figure 6.10(c)). This capability of the OR-PAM enables counting of petechiae (the rashes on the skin due to hemorrhage into the dermis) better than the gold standard tourniquet test. For the oxygen saturation experiments during arterial occlusion, six sets of volumetric data were acquired over 400s. After several steps of pre-processing such as skin-removal, laser-energy fluctuation normalization, high pass filtering for de-blurring and so on, spectral unmixing was performed. It was observed that the oxygen saturation decreased by ~5 min after the arterial occlusion. This observation was aligned with the normoxia and hypoxia studies in the animals. The system discussed here is very promising to image the microvasculature in the human finger at a very high resolution for functional information. Clinical correlation of the brachial cuff pressure experiments is yet to be validated. Studies must be done to observe the performance of the imaging system across different skin tones and age groups. Nonetheless, with further progress in the design of the system and clinical understanding, high-speed, high-resolution, multi-wavelength OR-PAM system can be a reliable screening and diagnostic tool in dermatology and cardiovascular fields.

6.6.2 Forearm imaging

The forearm is another imaging location for blood vessel studies (Bunke *et al* 2021). In this study done by Bunke *et al* seven healthy volunteers (3 men and 4 women, between 20 and 75 years old, skin type II–III on the Fitzpatrick scale) were imaged with PAI in the forearm to monitor localized vasoconstriction. A linear array-based system (Vevo LAZR-X instrument) was used for imaging. The center frequency of the transducer was 30 MHz. The PA images were acquired from 680 to 970 nm sweeping at an interval of 10 nm. 50 and 110 μm was the axial and lateral resolution of the system. The imaging probe was mounted on an adjustable arm to minimize the motion. Once the probe was positioned and the arm was secured, baseline PA images were acquired. Then, the subjects were injected with a concoction of 0.5 ml lidocaine (20 mg ml^{-1}) and adrenaline (12.5 μg ml^{-1}) in the dermis of forearm. This caused vasoconstriction of the blood vessels. Along with PA images, diffuse reflectance spectroscopy (DRS) measurements were also done, which provided HbO_2, HbT and HbR in arbitrary units and sO_2 as a percentage. Spectral unmixing was done using non-negative matrix factorization to extract the sO_2 from the PA images. Unmixing was done considering five contributing chromophores—HbO_2, HbR, fat, melanin, and water. From monitoring of PA images, it was observed that HbO_2 decreased with the increase in HbR, which was as expected due to the injection of a vasoconstrictor, which led to the gradual reduction in sO_2. The change in sO_2 was observed only in the dermis and the sO_2 remained unchanged in the hypodermis. This demonstrated that the PAI is capable of measuring sO_2. The rate of change of sO_2 calculated from PA and DRS was 123s and 109s, respectively. Based on the multiwavelength information of the PA images, spatially resolved maps of the sO_2 during adrenaline injection was feasible.

Heuman *et al* used a light-emitting diode (LED)-based PAI system to assist microsurgical lymphovenous bypass (LVB) which is the surgical treatment for secondary lymphedema (Van Heumen *et al* 2023). Fluorescence imaging is used as a gold standard to differentiate the healthy lymph vessels and the vessels that have dermal backflow. AcousticX system (figure 6.10(d)), composed of linear array transducer (center frequency 7 MHz and fractional bandwidth of 80%), and an LED emitting 820 and 940 nm was used for this study. 0.1 and 0.09 mJ cm^{-2} were the optical fluences on the skin. Different imaging orientations on the forearm are shown in figure 6.10(e). 64 frames averaging was done on DAQ and 10 frames averaging was done on the software leading to 640 averaging, which led to a dual-wavelength frame-rate of 6.25 Hz. Lateral and axial resolution of the system was 350 and 210 μm, respectively.

US images were acquired at a sampling rate of 20 MHz and PA images were acquired at a sampling rate of 40 MHz. Image processing was done to maintain the continuity of the blood vessels. The structures that had low 940/820 nm PA signal amplitude ratio was identified as lymphatic vessels. It was feasible to visualize lymph vessels in real-time at a resolution better than fluorescence imaging. In cases where there was dermal backflow, lymphatic vessels were differentiable from blood vessels in PA images, which is not feasible in fluorescence images. Figure 6.10(f) is the interlaid MIP of the 820 and 940 nm. The pink (absorber of both wavelengths) and the red arrows (absorber of 820 nm) are the blood vessel. For further clinical advancement of PA for lymph vessel imaging, the LED-based imaging system needs improvement in terms of artifact suppression and better 3D imaging.

6.6.3 Feet imaging

Imaging of human feet is crucial to understand the vascular health of diabetic patients (Choi *et al* 2022). Choi *et al* used the clinical PA/US system from E-CUBE 12R, Alpinion Medical Systems, with tunable pulsed laser that has a frame rate of 5 Hz to build a foot scanner. The 128-element clinical linear array probe was mounted on a three-axis motorized stage. The foot was secured inside the water using a footwear-shaped 3D printed structure, which also has an ankle support. Laser light was delivered to the target using optical fiber which was aligned with the ultrasound probe using a 3D printed holder. First, a US scan was done to map the topology of the foot. Next, simultaneous US and PA images were acquired. The images of the foot and the blood vessel network were acquired and the MAP images were computed. The limitation of this imaging system is that the deeper blood vessels cannot be imaged because of the bones.

Wang *et al* introduced a portable PAT using a linear array probe to study the vasculature of human foot with chronic ulcers (Wang *et al* 2019a). The 128-element linear array-based system was combined with 1064 nm from Nd:YAG laser. The illumination was at an angle of 45°, with respect to the transducer. The custom-made waterproof ultrasound transducer was 8.6 cm wide, had a center frequency of 2.25 MHz and a focal length of 4 cm. The lateral resolution was 0.7 mm and the elevational resolution was 1.3 mm. The blood vasculature in a healthy volunteer was imaged. Blood vessels up to 9 mm were seen in the PA images. The foot vasculature of two diabetes patients (41- and 63-year-olds) were also imaged. The blood vasculature was adequate according to the PA for the 41-year-old. The blood vasculature image of the 63-year-old exhibited poor vasculature indicating poor blood perfusion. This observation was validated with CT angiography. This patient was diagnosed with ischemic condition and vessel stenosis.

6.7 Next generation photoacoustic technology

The research on PAI has been looking towards advancing technology, biological parameter monitoring, and ultra-portable devices. This section will discuss some new applications of PAI in the monitoring space.

6.7.1 Augmented reality

Augmented reality and mixed reality have been finding their way into the clinical settings. This can improve the efficiency and risk of surgery. In developing augmented reality, PAI holds promise to provide users with a better visualization of the tissues that PA targets well, filling gaps that other visualization methods cannot image noninvasively or with high resolution. Pan *et al* created a system for real-time visualization of the deep microvascular network within the tissues on a surgical surface (Pan *et al* 2023). The system realized a vascular localization accuracy of <0.89 mm and a vascular relocation latency of <1 s. A visual tracking algorithm and a curved surface-fitting algorithm were developed to achieve this. The imaging system consists of an LED-based PA probe which took data simultaneously with a RGBD (R: Red, G: Green, B: Blue, D: Depth) camera. The PA data was reconstructed and sent to a projector. As the image was projected onto the surface, the RGBD tracked the target so that the preoperative images could be placed accurately on the physical surface.

6.7.2 Portable devices

Flexible sensing technologies have the potential to transform medical instruments and diagnostic methods. This type of technology provides diagnostic devices portability, wearability, and remote functionality. These characteristics could also aid in response time to medical conditions and events. The electronics themselves are based on non-invasive or minimally invasive portable sensors that capture biochemical and mechanical signals of the human body. The signals facilitate real-time monitoring of physiological parameters and encourage personalized medicine.

The rise of wearable sensing systems represents a significant advancement in healthcare technology, with the international market for these devices projected to soar from $20.1 billion in 2021 to $83.9 billion in 2026 (Butt *et al* 2022). This growth underscores the transformative potential of wearable medical sensing devices as next-generation portable remote healthcare tools. Furthermore, these systems play a crucial role in applications beyond healthcare, such as serving as skin sensory inputs for virtual reality (VR) and augmented reality (AR) systems, enhancing the human-machine interface (HMI) by transmitting vital information like temperature, pressure, and strain. Moreover, wearable sensing systems are integrated into smart prosthetics and soft robotics, augmenting their functionality through external signal-conversion, control, and execution modules. For these reasons, using PAI as a sensing and monitoring modality is an area of interest in the field of photoacoustics.

Zhang *et al* developed a wearable photoacoustic watch and backpack system for human hemoglobin monitoring (Zhang *et al* 2024). The system was fully integrated, and the watch had a size of 43 × 30 × 24 mm and a weight of 40 g. The sensing capability had a lateral resolution of 8.7 μm and a FOV of 3 mm in diameter. The focus was adjustable so that the device could be optimized via change of the imaging plane, to the wearer. While the watch/backpack system was being worn, imaging of the human wrist was done during various physiological activities. The feasibility of such a device, with customizable features, was established.

Gao *et al* proposed a wearable PA patch. Its function is to three-dimensionally map hemoglobin in deep tissue along with monitoring core temperature (Gao *et al* 2022). The patch was constructed on a soft substrate containing an array of US transducers and vertical-cavity surface-emitting laser (VCSEL) diodes. These diodes emitted a wavelength of 850 nm and had a penetration depth in tissue greater than 2 cm. Using this device the dimensions of blood vessels were able to be continuously measured *in vivo* though high-resolution imaging. The ability to monitor vessel function could help diagnose vascular diseases and indicate cardiac function during operations.

6.7.3 Medical monitoring

Using PAI technology to monitor physical parameters apart from wearable technology is another PA application. Uluç *et al* devised a method for non-invasive glucose monitoring using PA signal. In diabetic management and other clinical examinations, finger pricking is the common method by which blood glucose is measured (Uluç *et al* 2024). An accurate and less intrusive method would be beneficial to those affected. The sensor developed, termed depth-gated mid-infrared optoacoustic sensor (DIROS), was able to determine blood glucose levels more accurately than competing non-invasive monitoring methods. The sensor used combined mid-IR–visible (mid-IR/VIS) beam since glucose has strong absorption in this range. It also took advantage of time-gating to improve sensitivity and accuracy. They used a broadband US transducer that had a bandwidth of ~6–36 MHz and central frequency of ~21 MHz. The axial resolution in the upper frequency band was 30 μm. The experiment was performed *in vivo* on a mouse ear where depth sensitive glucose sensing was demonstrated. The Pearson correlation coefficient at a depth of 97.5 μm was $r = 0.92$ for and slightly lowered as the time gate moved shallower to $r = 0.\ 80$ and $r = 0.72$.

Padamanabhan *et al* investigated deep tissue sensing of chiral molecules using PA. Chirality is found in various organic and inorganic compounds (Padmanabhan and Prakash 2024). It can be an identifier of the presence of these compounds, which include proteins, amino acids, various sugars, lactates, and drugs such as ketamine, steroids, and beta-blockers. The current *in vivo* measurement method based on polarimetry has an imaging depth of only 1 mm. The proposed PA sensing method used signals at a NIR-II wavelengths of 1560 and 1500 nm to reduce scattering and enhance penetration depth. Experiments were performed *ex vivo* with chicken tissue and on serum-based samples. The results showed significant correlation in chiral molecule concentration in depths of up to 3.5 mm. It demonstrated a detection limit of 80 mg dl^{-1} using circular incidence polarization. This method shows promise for *in vivo* results as well as portability and scaling-down as it uses a single wavelength.

6.8 Conclusion

In conclusion, using PAI as a biomedical imaging modality has a strategic role in disease diagnosis and treatment. The applications of PAI have grown from pre-clinical small animal studies to clinical imaging in humans. The application areas are almost throughout the body. In pre-clinical and clinical settings, PAI has been used to image various organs and systems, including the brain, teeth, thyroid, epidermis, vasculature, cardiovascular system, breast, gynecological systems, gastro-intestinal tract, fingers, arms, feet, and others not included in this chapter. These applications showcase PAI's ability to provide detailed insights into physiological processes and pathologies. Ongoing research is exploring the potential of PAI in clinical settings, with promising applications in functional neuroimaging, breast imaging, and vascular assessment. In spite of the promises, there are several challenges one needs to address to make clinical translation in the future feasible (Assi *et al* 2023). Further, the way of the future in healthcare looks to be trending toward real-time continuous data gathering and interpretation of biomarkers. Medical monitoring, wearable, and patch devices are growing in popularity in preventative and supervisory health.

6.9 Problems

1. What are the advantages and disadvantages of various biomedical imaging modalities?
2. What are the challenges of pure optical imaging that PAI can alleviate?
3. Describe the photoacoustic effect.
4. What are the different types of PAI systems?
5. Compare photoacoustic computed tomography with photoacoustic microscopy.
6. What type of light sources are needed for PAI?
7. What kind of functional parameters can be measured/imaged with PAI.
8. Is there any safety precaution one needs to take for PAI?
9. Are there any challenges for photoacoustic brain imaging?
10. Do you see any challenges for full body imaging with PAI? What steps can be taken to improve the imaging depth?
11. What can be done to improve the internal organ imaging with PAI?
12. What application of PAI will have faster clinical translation? Justify your response.
13. Describe the challenges for making PAI as a portable device.
14. What kind of physiological parameters can be monitored using PAI?
15. What are the future directions of PAI?
16. What are innovations in other scientific fields that may impact/improve PAI in the future?

References

Acharya R, Wasserman R, Stevens J and Hinojosa C 1995 Biomedical imaging modalities: a tutorial *Comput. Med. Imaging Graph.* **19** 3–25

Agrawal S, Fadden C, Dangi A, Yang X, Albahrani H, Frings N, Heidari Zadi S and Kothapalli S R 2019 Light-emitting-diode-based multispectral photoacoustic computed tomography system *Sensors* **19** 4861

Ahn J, Kim J Y, Choi W and Kim C 2021 High-resolution functional photoacoustic monitoring of vascular dynamics in human fingers *Photoacoustics* **23** 100282

Ali Z *et al* 2022 360° optoacoustic capsule endoscopy at 50 Hz for esophageal imaging *Photoacoustics* **25** 100333

Amidi E, Yang G, Uddin K M S, Luo H B, Middleton W, Powell M, Siegel C and Zhu Q 2021 Role of blood oxygenation saturation in ovarian cancer diagnosis using multi-spectral photoacoustic tomography *J. Biophotonics* **14** e202000368

Andreotti R F *et al* 2020 O-RADS US Risk Stratification and Management System: a consensus guideline from the ACR Ovarian-Adnexal Reporting and Data System Committee *Radiology* **294** 168–85

Ansari R, Zhang E Z, Desjardins A E and Beard P C 2020 Miniature all-optical flexible forward-viewing photoacoustic endoscopy probe for surgical guidance *Opt. Lett.* **45** 6238–41

Arabul M U, Heres M, Rutten M C M, van Sambeek M R, Van de Vosse F N and Lopata R G P 2017 Toward the detection of intraplaque hemorrhage in carotid artery lesions using photoacoustic imaging *J. Biomed. Opt.* **22** ARTN 041010

Arnold M *et al* 2022 Current and future burden of breast cancer: global statistics for 2020 and 2040 *Breast* **66** 15–23

Assi H *et al* 2023 A review of strategic roadmapping exercise to advance clinical translation of photoacoustic imaging: from current barriers to future adoption *Photoacoustics* **32** 100539

Awasthi N, Kalva S K, Pramanik M and Yalavarthy P K 2021 Dimensionality reduced plug and play priors for improving photoacoustic tomographic imaging with limited noisy data *Biomed. Opt. Express* **12** 1320–38

Benson J R, Jatoi I, Keisch M, Esteva F J, Makris A and Jordan V C 2009 Early breast cancer *Lancet* **373** 1463–79

Brédart A, Kop J L, Fall M, Pelissier S, Simondi C, Dolbeault S, Livartowski A, Tardivon A and Study M R I 2012 Perception of care and experience of examination in women at risk of breast cancer undergoing intensive surveillance by standard imaging with or without MRI *Patient Educ. Couns.* **86** 405–13

Britton P *et al* 2009 One-stop diagnostic breast clinics: how often are breast cancers missed? *Br. J. Cancer* **100** 1873–8

Brunker J, Yao J, Laufer J and Bohndiek S E 2017 Photoacoustic imaging using genetically encoded reporters: a review *J. Biomed. Opt.* **22** 070901

Bunke J, Merdasa A, Sheikh R, Albinsson J, Erlov T, Gesslein B, Cinthio M, Reistad N and Malmsjo M 2021 Photoacoustic imaging for the monitoring of local changes in oxygen saturation following an adrenaline injection in human forearm skin *Biomed. Opt. Express* **12** 4084–96

Butt M A, Kazanskiy N L and Khonina S N 2022 Revolution in flexible wearable electronics for temperature and pressure monitoring—a review *Electronics* **11** 716

Carmeliet P and Jain R K 2011 Molecular mechanisms and clinical applications of angiogenesis *Nature* **473** 298–307

Chatni M R *et al* 2012 Tumor glucose metabolism imaged *in vivo* in small animals with whole-body photoacoustic computed tomography *J. Biomed. Opt.* **17** 076012

Chen J, Liu C, Hu D, Wang F, Wu H, Gong X, Liu X, Song L, Sheng Z and Zheng H 2016 Single-layer MoS_2 nanosheets with amplified photoacoustic effect for highly sensitive photoacoustic imaging of orthotopic brain tumors *Adv. Funct. Mater.* **26** 8715–25

Cheng R, Shao J, Gao X, Tao C, Ge J and Liu X 2016 Noninvasive assessment of early dental lesion using a dual-contrast photoacoustic tomography *Sci. Rep.* **6** 21798

Choi W *et al* 2022 Three-dimensional multistructural quantitative photoacoustic and US imaging of human feet *in vivo Radiology* **303** 467–73

Chuah S Y, Attia A B E, Ho C J H, Li X, Lee J S-S, Tan M W P, Yong A A, Tan A W M, Razansky D and Olivo M 2019 Volumetric multispectral optoacoustic tomography for 3-dimensional reconstruction of skin tumors: a further evaluation with histopathologic correlation *J. Invest. Dermatol.* **139** 481–5

Chuangsuwanich T, Moothanchery M, Yan A T C, Schmetterer L, Girard M J A and Pramanik M 2018 Photoacoustic imaging of lamina cribrosa microcapillary in porcine eyes *Appl. Opt.* **57** 4865–71

Dang X, Bardhan N M, Qi J, Gu L, Eze N A, Lin C-W, Kataria S, Hammond P T and Belcher A M 2019 Deep-tissue optical imaging of near cellular-sized features *Sci. Rep.* **9** 3873

Das D, Sharma A, Rajendran P and Pramanik M 2021 Another decade of photoacoustic imaging *Phys. Med. Biol.* **66** 05TR01

Das D, Sivasubramanian K, Yang C and Pramanik M 2018 On-chip generation of microbubbles in photoacoustic contrast agents for dual modal ultrasound/photoacoustic *in vivo* animal imaging *Sci. Rep.* **8** 6401

Deán-Ben X and Razansky D 2019 Optoacoustic image formation approaches—a clinical perspective *Phys. Med. Biol.* **64** 18TR01

Dima A and Ntziachristos V 2016 In-vivo handheld optoacoustic tomography of the human thyroid *Photoacoustics* **4** 65–9

El-Sharkawy Y H and El Sherif A F 2012 Photoacoustic diagnosis of human teeth using interferometric detection scheme *Opt. Laser Technol.* **44** 1501–6

Emilio M D P 2019 *High-Performance Design for Ultrasound Imaging* https://eetimes.com/high-performance-design-for-ultrasound-imaging/

Essink-Bot M-L, Rijnsburger A, van Dooren S, De Koning H and Seynaeve C 2006 Women's acceptance of MRI in breast cancer surveillance because of a familial or genetic predisposition *Breast* **15** 673–6

Estrada H, Turner J, Kneipp M and Razansky D 2014 Real-time optoacoustic brain microscopy with hybrid optical and acoustic resolution *Laser Phys. Lett.* **11** 045601

Favazza C P, Cornelius L A and Wang L V 2011 In vivo functional photoacoustic microscopy of cutaneous microvasculature in human skin *J. Biomed. Opt.* **16** 026004

Fu L *et al* 2022 Photoacoustic imaging of posterior periodontal pocket using a commercial hockey-stick transducer *J. Biomed. Opt.* **27** 056005

Gangwani M K, Aziz A, Dahiya D S, Nawras M, Aziz M and Inamdar S 2023 History of colonoscopy and technological advances: a narrative review *Transl. Gastroenterol. Hepatol.* **8** 18

Gao C, Deng Z-J, Peng D, Jin Y-S, Ma Y, Li Y-Y, Zhu Y-K, Xi J-Z, Tian J and Dai Z-F 2016 Near-infrared dye-loaded magnetic nanoparticles as photoacoustic contrast agent for enhanced tumor imaging *Cancer Biol. Med.* **13** 349–59

Gao P, Chen Z Q, Bao Y H, Jiao L Q and Ling F 2007 Correlation between carotid intraplaque hemorrhage and clinical symptoms—systematic review of observational studies *Stroke* **38** 2382–90

Gao X *et al* 2022 A photoacoustic patch for three-dimensional imaging of hemoglobin and core temperature *Nat. Commun.* **13** 7757

Gao X H, Li W, Loomes M and Wang L Y 2017 A fused deep learning architecture for viewpoint classification of echocardiography *Inf. Fusion* **36** 103–13

Grinvald A, Frostig R, Lieke E and Hildesheim R 1988 Optical imaging of neuronal activity *Physiol. Rev.* **68** 1285–366

Gu Y, Sun Y, Wang X, Li H, Qiu J and Lu W 2023 Application of photoacoustic computed tomography in biomedical imaging: a literature review *Bioeng. Transl. Med.* **8** e10419

Guo B, Sheng Z, Hu D, Lin X, Xu S, Liu C, Zheng H and Liu B 2017 Biocompatible conjugated polymer nanoparticles for highly efficient photoacoustic imaging of orthotopic brain tumors in the second near-infrared window *Mater. Horiz.* **4** 1151–6

Guo H, Li Y, Qi W and Xi L 2020 Photoacoustic endoscopy: a progress review *J. Biophotonics* **13** e202000217

Gutta S, Kalva S K, Pramanik M and Yalavarthy P K 2018 Accelerated image reconstruction using extrapolated Tikhonov filtering for photoacoustic tomography *Med. Phys.* **45** 3749–67

Haindl R *et al* 2023 Visible light photoacoustic ophthalmoscopy and near-infrared-II optical coherence tomography in the mouse eye *APL Photonics* **8** 106108

Han X *et al* 2014 Endoscopic biopsy in gastrointestinal neuroendocrine neoplasms: a retrospective study *PLoS One* **9** e103210

Hanahan D and Weinberg R A 2011 Hallmarks of cancer: the next generation *Cell* **144** 646–74

He H L, Buehler A, Bozhko D, Jian X H, Cui Y Y and Ntziachristos V 2018 Importance of ultrawide bandwidth for optoacoustic esophagus imaging *IEEE Trans. Med. Imaging* **37** 1162–7

He H L, Englert L and Ntziachristos V 2023 Optoacoustic endoscopy of the gastrointestinal tract *ACS Photonics* **10** 559–70

He H L, Stylogiannis A, Afshari P, Wiedemann T, Steiger K, Buehler A, Zakian C and Ntziachristos V 2019 Capsule optoacoustic endoscopy for esophageal imaging *J. Biophotonics* **12** e201800439

Heijblom M, Piras D, Brinkhuis M, van Hespen J C, van den Engh F M, van der Schaaf M, Klaase J M, van Leeuwen T G, Steenbergen W and Manohar S 2015 Photoacoustic image patterns of breast carcinoma and comparisons with magnetic resonance imaging and vascular stained histopathology *Sci. Rep.* **5** 11778

Hennen S N, Xing W, Shui Y B, Zhou Y, Kalishman J, Andrews-Kaminsky L B, Kass M A, Beebe D C, Maslov K I and Wang L V 2015 Photoacoustic tomography imaging and estimation of oxygen saturation of hemoglobin in ocular tissue of rabbits *Exp. Eye Res.* **138** 153–8

Hoge R D, Atkinson J, Gill B, Crelier G R, Marrett S and Pike G B 1999 Investigation of BOLD signal dependence on cerebral blood flow and oxygen consumption: the deoxyhemoglobin dilution model *Magn. Reson. Med.* **42** 849–63

Hooley R J, Andrejeva L and Scoutt L M 2011 Breast cancer screening and problem solving using mammography, ultrasound, and magnetic resonance imaging *Ultrasound Q.* **27** 23–47

Huang C, Nie L, Schoonover R W, Guo Z, Schirra C O, Anastasio M A and Wang L V 2012 Aberration correction for transcranial photoacoustic tomography of primates employing adjunct image data *J. Biomed. Opt.* **17** 1–8 8

Huang S, Upputuri P K, Liu H, Pramanik M and Wang M 2016 A dual-functional benzobis-thiadiazole derivative as an effective theranostic agent for near-infrared photoacoustic imaging and photothermal therapy *J. Mater. Chem.* B **4** 1696–703

Hughes D, Sampathkumar A, Longbottom C and Kirk K 2015 Imaging and detection of early stage dental caries with an all-optical photoacoustic microscope *J. Phys.: Conf. Ser.* **581** 012002

Hui X, Mohammad M O A and Pramanik M 2022 Looking deep inside tissue with photoacoustic molecular probes: a review *J. Biomed. Opt.* **27** 070901

Hui X, Rajendran P, Zulkifli M A I, Ling T and Pramanik M 2023 Android mobile-platform-based image reconstruction for photoacoustic tomography *J. Biomed. Opt.* **28** 046009

Ikematsu H *et al* 2022 Photoacoustic imaging of fresh human surgically and endoscopically resected gastrointestinal specimens *Den Open.* **2** e28

Jansen K, van der Steen A F W, van Beusekom H M M, Oosterhuis J W and van Soest G 2011 Intravascular photoacoustic imaging of human coronary atherosclerosis *Opt. Lett.* **36** 597–9

Jansen K, van Soest G and van der Steen A F W 2014 Intravascular photoacoustic imaging: a new tool for vulnerable plaque identification *Ultrasound Med. Biol.* **40** 1037–48

Jeon S, Kim J, Lee D, Baik J W and Kim C 2019 Review on practical photoacoustic microscopy *Photoacoustics* **15** 100141

Jo J, Xu G, Cao M, Marquardt A, Francis S, Gandikota G and Wang X 2017 A functional study of human inflammatory arthritis using photoacoustic imaging *Sci. Rep.* **7** 15026

Kalva S K, Hui Z Z and Pramanik M 2018 Calibrating reconstruction radius in a multi single-element ultrasound-transducer-based photoacoustic computed tomography system *J. Opt. Soc. Am.* A **35** 764–71

Kalva S K, Upputuri P K, Rajendran P, Dienzo R A and Pramanik M 2019 Pulsed laser diode-based desktop photoacoustic tomography for monitoring wash-in and wash-out of dye in rat cortical vasculature *J. Vis. Exp.* **30** e59764

Kenry, Duan Y and Liu B 2018 Recent advances of optical imaging in the second near-infrared window *Adv. Mater.* **30** 1802394

Kim C, Favazza C and Wang L V 2010 In vivo photoacoustic tomography of chemicals: high-resolution functional and molecular optical imaging at new depths *Chem. Rev.* **110** 2756–82

Kobayashi H *et al* 2008 A randomized study of screening for ovarian cancer: a multicenter study in Japan *Int. J. Gynecol. Cancer* **18** 414–20

Kopans D B 2007 *Breast Imaging* (Philadelphia, PA: Lippincott Williams & Wilkins)

Korenbrot J I 2012 Speed, sensitivity, and stability of the light response in rod and cone photoreceptors: facts and models *Prog. Retin. Eye Res.* **31** 442–66

Kratkiewicz K, Manwar R, Zhou Y, Mozaffarzadeh M and Avanaki K 2021 Technical considerations in the verasonics research ultrasound platform for developing a photoacoustic imaging system *Biomed. Opt. Express* **12** 1050–84

Ku G, Zhou M, Song S, Huang Q, Hazle J and Li C 2012 Copper sulfide nanoparticles as a new class of photoacoustic contrast agent for deep tissue imaging at 1064 nm *ACS Nano* **6** 7489–96

Kuhl C K 2007 Current status of breast MR imaging—part 2. Clinical applications *Radiology* **244** 672–91

Lashof J C, Henderson I C and Nass S J 2001 *Mammography and Beyond: Developing Technologies for the Early Detection of Breast Cancer* (National Academies Press)

Lee D, Park S, Noh W-C, Im J-S and Kim C 2017 Photoacoustic imaging of dental implants in a porcine jawbone ex vivo *Opt. Lett.* **42** 1760–3

Lee H, Han S, Kye H, Kim T-K, Choi W and Kim J 2023 A review on the roles of photoacoustic imaging for conventional and novel clinical diagnostic applications *Photonics* **10** 904

Lee H Y, Park E C, Jun J K, Choi K S and Hahm M I 2010 Comparing upper gastrointestinal x-ray and endoscopy for gastric cancer diagnosis in Korea *World J. Gastroenterol.* **16** 245–50

Li L *et al* 2017 Single-impulse panoramic photoacoustic computed tomography of small-animal whole-body dynamics at high spatiotemporal resolution *Nat. Biomed. Eng.* **1** 0071

Li Y M, Suki D, Hess K and Sawaya R 2016 The influence of maximum safe resection of glioblastoma on survival in 1229 patients: can we do better than gross-total resection? *J. Neurosurg.* **124** 977–88

Liang Y Z, Fu W B, Li Q, Chen X L, Sun H J, Wang L D, Jin L, Huang W and Guan B O 2022 Optical-resolution functional gastrointestinal photoacoustic endoscopy based on optical heterodyne detection of ultrasound *Nat. Commun.* **13** 7604

Lin L, Hu P, Shi J, Appleton C M, Maslov K, Li L, Zhang R and Wang L V 2018 Single-breath-hold photoacoustic computed tomography of the breast *Nat. Commun.* **9** 2352

Lin L, Hu P, Tong X, Na S, Cao R, Yuan X, Garrett D C, Shi J, Maslov K and Wang L V 2021 High-speed three-dimensional photoacoustic computed tomography for preclinical research and clinical translation *Nat. Commun.* **12** 882

Lin L, Yao J, Li L and Wang L V 2016 In vivo photoacoustic tomography of myoglobin oxygen saturation *J. Biomed. Opt.* **21** 61002

Liu C *et al* 2019a In vivo transrectal imaging of canine prostate with a sensitive and compact handheld transrectal array photoacoustic probe for early diagnosis of prostate cancer *Biomed. Opt. Express* **10** 1707–17

Liu L, Chen Q, Wen L, Li C, Qin H and Xing D 2019b Photoacoustic therapy for precise eradication of glioblastoma with a tumor site blood–brain barrier permeability upregulating nanoparticle *Adv. Funct. Mater.* **29** 1808601

Liu W and Zhang H F 2016 Photoacoustic imaging of the eye: a mini review *Photoacoustics* **4** 112–23

Liu Y, Lv X, Liu H, Zhou Z, Huang J, Lei S, Cai S, Chen Z, Guo Y and Chen Z 2018 Porous gold nanocluster-decorated manganese monoxide nanocomposites for microenvironment-activatable MR/photoacoustic/CT tumor imaging *Nanoscale* **10** 3631–8

Lnsis A 2000 Atherosclenrosis *Nature* **407** 233–41

Lu W *et al* 2010 Photoacoustic imaging of living mouse brain vasculature using hollow gold nanospheres *Biomaterials* **31** 2617–26

Luke G P, Bashyam A, Homan K A, Makhija S, Chen Y S and Emelianov S Y 2013 Silica-coated gold nanoplates as stable photoacoustic contrast agents for sentinel lymph node imaging *Nanotechnology* **24** 455101

Maccioni F, Busato L, Valenti A, Cardaccio S, Longhi A and Catalano C 2023 Magnetic resonance imaging of the gastrointestinal tract: current role, recent advancements and future prospectives *Diagnostics (Basel)* **13** 2410

Macri F, Khasanova E, Niu B T, Parakh A, Patino M, Kambadakone A and Sahani D V 2023 Optimal abdominal CT image quality in non-lean patients: customization of cm injection protocols and low-energy acquisitions *Diagnostics (Basel)* **13** 2279

Madani A, Arnaout R, Mofrad M and Arnaout R 2018 Fast and accurate view classification of echocardiograms using deep learning *Npj Digit. Med.* **1** 6

Mannath J and Ragunath K 2016 Role of endoscopy in early oesophageal cancer *Nat. Rev. Gastroenterol. Hepatol.* **13** 720–30

Manohar S and Dantuma M 2019 Current and future trends in photoacoustic breast imaging *Photoacoustics* **16** 100134

Manohar S, Kharine A, van Hespen J C, Steenbergen W and van Leeuwen T G 2005 The Twente Photoacoustic Mammoscope: system overview and performance *Phys. Med. Biol.* **50** 2543

Manohar S and Razansky D 2016 Photoacoustics: a historical review *Adv. Opt. Photonics* **8** 586–617

Maslov K, Stoica G and Wang L V 2005 *In vivo* dark-field reflection-mode photoacoustic microscopy *Opt. Lett.* **30** 625–7

McDonald R J, Levine D, Weinreb J, Kanal E, Davenport M S, Ellis J H, Jacobs P M, Lenkinski R E, Maravilla K R and Prince M R 2018 Gadolinium retention: a research roadmap from the 2018 NIH/ACR/RSNA workshop on gadolinium chelates *Radiology* **289** 517–34

Miao Q and Pu K 2018 Organic semiconducting agents for deep-tissue molecular imaging: second near-infrared fluorescence, self-luminescence, and photoacoustics *Adv. Mater.* **30** 1801778

Moore C, Cheng Y, Tjokro N, Zhang B, Kerr M, Hayati M, Chang K C J, Shah N, Chen C and Jokerst J V 2022 A photoacoustic-fluorescent imaging probe for proteolytic gingipains expressed by porphyromonas gingivalis *Angew Chem. Int. Ed. Engl.* **61** e202201843

Moothanchery M, Dev K, Balasundaram G, Bi R and Olivo M 2019 Acoustic resolution photoacoustic microscopy based on MEMS scanner *J. Biophotonics* **13** e201960127

Moothanchery M and Pramanik M 2017 Performance characterization of a switchable acoustic and optical resolution photoacoustic microscopy system *Sensors* **17** 357

Muller J W, van Hees R, van Sambeek M, Boutouyrie P, Rutten M, Brands P, Wu M and Lopata R 2021 Towards *in vivo* photoacoustic imaging of vulnerable plaques in the carotid artery *Biomed. Opt. Express* **12** 4207–18

Na S *et al* 2021 Massively parallel functional photoacoustic computed tomography of the human brain *Nat. Biomed. Eng.* **6** 584–92

Na S and Wang L V 2021 Photoacoustic computed tomography for functional human brain imaging [Invited] *Biomed. Opt. Express* **12** 4056–83

Naghavi M and Falk E 2010 From vulnerable plaque to vulnerable patient *Asymptomatic Atherosclerosis: Pathophysiology, Detection and Treatment* (Berlin: Springer) pp 13–38

Nandy S, Mostafa A, Hagemann I S, Powell M A, Amidi E, Robinson K, Mutch D G, Siegel C and Zhu Q 2018 Evaluation of ovarian cancer: initial application of coregistered photoacoustic tomography and US *Radiology* **289** 740–7

Ni D, Zhang J, Bu W, Xing H, Han F, Xiao Q, Yao Z, Chen F, He Q and Liu J 2014 Dual-targeting upconversion nanoprobes across the blood–brain barrier for magnetic resonance/fluorescence imaging of intracranial glioblastoma *ACS Nano* **8** 1231–42

Nie H L, Luo H B, Chen L and Zhu Q 2023 A coregistered ultrasound and photoacoustic imaging protocol for the transvaginal imaging of ovarian lesions *Jove-J. Vis. Exp.* e64864

Nishiyama M, Namita T, Kondo K, Yamakawa M and Shiina T 2019 Ring-array photoacoustic tomography for imaging human finger vasculature *J. Biomed. Opt.* **24** 096005

Onesti J K, Mangus B E, Helmer S D and Osland J S 2008 Breast cancer tumor size: correlation between magnetic resonance imaging and pathology measurements *Am. J. Surg.* **196** 844–8

Ono H *et al* 2021 Guidelines for endoscopic submucosal dissection and endoscopic mucosal resection for early gastric cancer (second edition) *Digest. Endosc.* **33** 4–20

Oraevsky A A, Jacques S L, Esenaliev R O and Tittel F K 1994 Laser-based optoacoustic imaging in biological tissues *Proc. of Laser-Tissue Interaction V* **2134** 122–8

Oraevsky A A, Karabutov A A, Solomatin S V, Savateeva E V, Andreev V A, Gatalica Z, Singh H and Fleming R D 2001 Laser optoacoustic imaging of breast cancer *in vivo Biomedical Optoacoustics II: SPIE* (Bellingham, WA: SPIE) 6–15

Padmanabhan S and Prakash J 2024 Deep tissue sensing of chiral molecules using polarization enhanced photoacoustics arXiv preprint arXiv:2401.10812.

Pan S *et al* 2023 Photoacoustic-enabled automatic vascular navigation: accurate and naked-eye real-time visualization of deep-seated vessels *Adv. Photonics Nex.* **2** 046001

Park S, Lee C, Kim J and Kim C 2014 Acoustic resolution photoacoustic microscopy *Biomed. Eng. Lett.* **4** 213–22

Park S, Park G, Kim J, Choi W, Jeong U and Kim C 2018 Bi_2Se_3 nanoplates for contrast-enhanced photoacoustic imaging at 1064 nm *Nanoscale* **10** 20548–58

Parray F Q, Wani M L, Bijli A H, Thakur N, Irshad I and Nayeem ul H 2011 Crohn's disease: a surgeon's perspective *Saudi J. Gastroenterol.* **17** 6–15

Pediconi F, Catalano C, Roselli A, Dominelli V, Cagioli S, Karatasiou A, Pronio A, Kirchin M A and Passariello R 2009 The challenge of imaging dense breast parenchyma? *Invest. Radiol.* **44** 412–21

Periyasamy V, Das N, Sharma A and Pramanik M 2019 1064 nm acoustic resolution photoacoustic microscopy *J. Biophotonics* **12** e201800357

Periyasamy V, Gisi K and Pramanik M 2024 Principles and applications of photoacoustic computed tomography ed W Xia *Biomedical Photoacoustics: Technology and Applications* (Cham: Springer Nature) pp 75–107

Poudel J, Na S, Wang L V and Anastasio M A 2020 Iterative image reconstruction in transcranial photoacoustic tomography based on the elastic wave equation *Phys. Med. Biol.* **65** 055009

Prakash J, Sanny D, Kalva S K, Pramanik M and Yalavarthy P K 2019 Fractional regularization to improve photoacoustic tomographic image reconstruction *IEEE Trans. Med. Imaging* **38** 1935–47

Pramanik M 2014 Improving tangential resolution with a modified delay-and-sum reconstruction algorithm in photoacoustic and thermoacoustic tomography *J. Opt. Soc. Am.* A **31** 621–7

Pramanik M, Swierczewska M, Green D, Sitharaman B and Wang L V 2009 Single-walled carbon nanotubes as a multimodal-thermoacoustic and photoacoustic-contrast agent *J. Biomed. Opt.* **14** 034018

Qiu T, Lan Y, Gao W, Zhou M, Liu S, Huang W, Zeng S, Pathak J L, Yang B and Zhang J 2021a Photoacoustic imaging as a highly efficient and precise imaging strategy for the evaluation of brain diseases *Quant. Imaging Med. Surg.* **11** 2169

Qiu T, Lan Y T, Gao W J, Zhou M Y, Liu S Q, Huang W Y, Zeng S J, Pathak J L, Yang B and Zhang J 2021b Photoacoustic imaging as a highly efficient and precise imaging strategy for the evaluation of brain diseases *Quant. Imaging Med. Surg.* **11** 2169–86

Samanta D, Paul S, Paramanick A, Raval V R and Suheshkumar Singh M 2023 High-resolution imaging of the whole eye with photoacoustic microscopy *Opt. Lett.* **48** 3443–6

Schoeder S, Olefir I, Kronbichler M, Ntziachristos V and Wall W A 2018 Optoacoustic image reconstruction: the full inverse problem with variable bases *Proc. Royal Soc. A: Math., Phys. Eng. Sci.* **474** 20180369

Schwarz M, Buehler A, Aguirre J and Ntziachristos V 2016 Three-dimensional multispectral optoacoustic mesoscopy reveals melanin and blood oxygenation in human skin *in vivo J. Biophotonics* **9** 55–60

Seyedian S S, Nokhostin F and Malamir M D 2019 A review of the diagnosis, prevention, and treatment methods of inflammatory bowel disease *J. Med. Life* **12** 113–22

Shabairou N, Lengenfelder B, Hohmann M, Klampfl F, Schmidt M and Zalevsky Z 2020 All-optical, an ultra-thin endoscopic photoacoustic sensor using multi-mode fiber *Sci. Rep.* **10** 9142

Shiina T, Toi M and Yagi T 2018 Development and clinical translation of photoacoustic mammography *Biomed. Eng. Lett.* **8** 157–65

Shishikura D 2016 Noninvasive imaging modalities to visualize atherosclerotic plaques *Cardiovas. Diagn. Ther.* **6** 340–53

Singh M K A, Sato N, Ichihashi F and Sankai Y 2020 Clinical translation of photoacoustic imaging—opportunities and challenges from an industry perspective ed M Kuniyil Ajith Singh *LED-Based Photoacoustic Imaging : From Bench to Bedside* (Singapore: Springer) 379–93

Sivasubramanian K, Mathiyazhakan M, Wiraja C, Upputuri P K, Xu C and Pramanik M 2017a Near Infrared light-responsive liposomal contrast agent for photoacoustic imaging and drug release applications *J. Biomed. Opt.* **22** 041007

Sivasubramanian K, Periyasamy V and Pramanik M 2017b Hand-held clinical photoacoustic imaging system for real-time non-invasive small animal imaging *J. Vis. Exp.* **128** e56649

Sivasubramanian K, Periyasamy V and Pramanik M 2018 Non-invasive sentinel lymph node mapping and needle guidance using clinical handheld photoacoustic imaging system in small animal *J. Biophotonics* **11** e201700061

Sivasubramanian K, Periyasamy V, Wen K K and Pramanik M 2017c Optimizing light delivery through fiber bundle in photoacoustic imaging with clinical ultrasound system: Monte Carlo simulation and experimental validation *J. Biomed. Opt.* **22** 041008

Song L, Maslov K, Shung K K and Wang L V 2010 Ultrasound-array-based real-time photo-acoustic microscopy of human pulsatile dynamics *in vivo J. Biomed. Opt.* **15** 021303–021303-4

Tang J, Coleman J E, Dai X and Jiang H 2016 Wearable 3-D photoacoustic tomography for functional brain imaging in behaving rats *Sci. Rep.* **6** 25470

Tasmara F A, Widyaningrum R, Setiawan A and Mitrayana M 2023 Photoacoustic imaging of hidden dental caries using visible-light diode laser *J. Appl. Clin. Med. Phys.* **24** e13935

Teh J L, Shabbir A, Yuen S and So J B 2020 Recent advances in diagnostic upper endoscopy *World J. Gastroenterol.* **26** 433–47

Temkin S M, Miller E A, Samimi G, Berg C D, Pinsky P and Minasian L 2017 Outcomes from ovarian cancer screening in the PLCO trial: histologic heterogeneity impacts detection, overdiagnosis and survival *Eur. J. Cancer* **87** 182–8

Tian C, Zhang W, Mordovanakis A, Wang X and Paulus Y M 2017 Noninvasive chorioretinal imaging in living rabbits using integrated photoacoustic microscopy and optical coherence tomography *Opt. Express* **25** 15947–55

Treeby B E and Cox B T 2010 k-Wave: MATLAB toolbox for the simulation and reconstruction of photoacoustic wave fields *J. Biomed. Opt.* **15** 021314

Uluç N, Glasl S, Gasparin F, Yuan T, He H, Jüstel D, Pleitez M A and Ntziachristos V 2024 Non-invasive measurements of blood glucose levels by time-gating mid-infrared optoacoustic signals *Nat. Metab.* **6** 678–86

Upputuri P K and Pramanik M 2015 Performance characterization of low-cost, high-speed, portable pulsed laser diode photoacoustic tomography (PLD-PAT) system *Biomed. Opt. Express* **6** 4118–29

Upputuri P K and Pramanik M 2017 Recent advances toward preclinical and clinical translation of photoacoustic tomography: a review *J. Biomed. Opt.* **22** 041006

Upputuri P K and Pramanik M 2019 Photoacoustic imaging in the second near-infrared window: a review *J. Biomed. Opt.* **24** 040901

Upputuri P K and Pramanik M 2020 Recent advances in photoacoustic contrast agents for *in vivo* imaging *Wiley Interdiscip. Rev. Nanomed. Nanobiotechnol.* **12** e1618

Vaccarella S, Franceschi S, Bray F, Wild C P, Plummer M and Dal Maso L 2016 Worldwide thyroid-cancer epidemic? The increasing impact of overdiagnosis *N. Engl. J. Med.* **375** 614–7

van den Berg P J, Daoudi K, Moens H J B and Steenbergen W 2017 Feasibility of photoacoustic/ultrasound imaging of synovitis in finger joints using a point-of-care system *Photoacoustics* **8** 8–14

Van Heumen S, Riksen J J M, Singh M K A, Van Soest G and Vasilic D 2023 LED-based photoacoustic imaging for preoperative visualization of lymphatic vessels in patients with secondary limb lymphedema *Photoacoustics* **29** 100446

Vinnicombe S, Pinto Pereira S M, McCormack V A, Shiel S, Perry N and Silva I M D 2009 Full-field digital versus screen-film mammography: comparison within the UK breast screening program and systematic review of published data *Radiology* **251** 347–58

Wang B, Xiang L, Jiang M S, Yang J, Zhang Q, Carney P R and Jiang H 2012a Photoacoustic tomography system for noninvasive real-time three-dimensional imaging of epilepsy *Biomed. Opt. Express* **3** 1427–32

Wang B, Xiao J and Jiang H 2014 Simultaneous real-time 3D photoacoustic tomography and EEG for neurovascular coupling study in an animal model of epilepsy *J. Neural Eng.* **11** 046013

Wang F *et al* 2023 Global, regional, and national burden of digestive diseases: findings from the global burden of disease study 2019 *Front. Public Health* **11** 1202980

Wang K, Li C, Chen R and Shi J 2021 Recent advances in high-speed photoacoustic microscopy *Photoacoustics* **24** 100294

Wang P, Wang P, Wang H W and Cheng J X 2012b Mapping lipid and collagen by multispectral photoacoustic imaging of chemical bond vibration *J. Biomed. Opt.* **17** 96010–1

Wang P H, Liu H L, Hsu P H, Lin C Y, Wang C R, Chen P Y, Wei K C, Yen T C and Li M L 2012c Gold-nanorod contrast-enhanced photoacoustic micro-imaging of focused-ultrasound induced blood-brain-barrier opening in a rat model *J. Biomed. Opt.* **17** 061222

Wang Y, Zhan Y, Harris L M, Khan S and Xia J 2019a A portable three-dimensional photoacoustic tomography system for imaging of chronic foot ulcers *Quant. Imaging Med. Surg.* **9** 799

Wang Z *et al* 2019b pH-sensitive and biodegradable charge-transfer nanocomplex for second near-infrared photoacoustic tumor imaging *Nano Res.* **12** 49–55

Wang Z, Zhen X, Upputuri P K, Jiang Y, Lau J, Pramanik M, Pu K and Xing B 2019c Redox-activatable and acid-enhanced nanotheranostics for second near-infrared photoacoustic tomography and combined photothermal tumor therapy *ACS Nano* **13** 5816–25

Windra Sari A, Widyaningrum R, Setiawan A and Mitrayana 2023 Recent development of photoacoustic imaging in dentistry: a review on studies over the last decade *Saudi Dental J.* **35** 423–36

Wu M, Awasthi N, Rad N M, Pluim J P W and Lopata R G P 2021 Advanced ultrasound and photoacoustic imaging in cardiology *Sensors* **21** 9747

Wu M, Jansen K, Springeling G, van der Steen A F W and van Soest G 2014 Impact of device geometry on the imaging characteristics of an intravascular photoacoustic catheter *Appl. Opt.* **53** 8131–9

Wu M, van der Steen A F, Regar E and van Soest G 2016 Emerging technology update intravascular photoacoustic imaging of vulnerable atherosclerotic plaque *Interv. Cardiol. Rev.* **11** 120

Xiang L, Ji L, Zhang T, Wang B, Yang J, Zhang Q, Jiang M S, Zhou J, Carney P R and Jiang H 2013 Noninvasive real time tomographic imaging of epileptic foci and networks *Neuroimage* **66** 240–8

Xu G, Rajian J R, Girish G, Kaplan M J, Fowlkes J B, Carson P L and Wang X 2013 Photoacoustic and ultrasound dual-modality imaging of human peripheral joints *J. Biomed. Opt.* **18** 010502

Xu M and Wang L V 2006 Photoacoustic imaging in biomedicine *Rev. Sci. Instrum.* **77** 041101

Yang G, Amidi E and Zhu Q 2021 Photoacoustic tomography reconstruction using lag-based delay multiply and sum with a coherence factor improves *in vivo* ovarian cancer diagnosis *Biomed. Opt. Express* **12** 2250–63

Yang J M, Maslov K, Yang H C, Zhou Q F, Shung K K and Wang L V 2009a Photoacoustic endoscopy *Opt. Lett.* **34** 1591–3

Yang M, Zhao L, He X, Su N, Zhao C, Tang H, Hong T, Li W, Yang F and Lin L 2017 Photoacoustic/ultrasound dual imaging of human thyroid cancers: an initial clinical study *Biomed. Opt. Express* **8** 3449–57

Yang X, Maurudis A, Gamelin J, Aguirre A, Zhu Q and Wang L V 2009b Photoacoustic tomography of small animal brain with a curved array transducer *J. Biomed. Opt.* **14** 054007

Yang X and Wang L V 2008 Monkey brain cortex imaging by photoacoustic tomography *J. Biomed. Opt.* **13** 044009

Yao D K, Maslov K, Shung K K, Zhou Q and Wang L V 2010 In vivo label-free photoacoustic microscopy of cell nuclei by excitation of DNA and RNA *Opt. Lett.* **35** 4139–41

Yao J J, Xia J, Maslov K I, Nasiriavanaki M, Tsytsarev V, Demchenko A V and Wang L V 2013 Noninvasive photoacoustic computed tomography of mouse brain metabolism *in vivo* *Neuroimage* **64** 257–66

Yu J, Nguyen H N Y, Steenbergen W and Kim K 2018 Recent development of technology and application of photoacoustic molecular imaging toward clinical translation *J. Nucl. Med.* **59** 1202–7

Zhang D *et al* 2018 Self-quenched metal-organic particles as dual-mode therapeutic agents for photoacoustic imaging-guided second near-infrared window photochemotherapy *ACS Appl. Mater. Interfaces* **10** 25203–12

Zhang H F, Maslov K, Stoica G and Wang L V 2006 Functional photoacoustic microscopy for high-resolution and noninvasive *in vivo* imaging *Nat. Biotechnol.* **24** 848–51

Zhang J, Zhen X, Upputuri P K, Pramanik M, Chen P and Pu K 2017 Activatable photoacoustic nanoprobes for *in vivo* ratiometric imaging of peroxynitrite *Adv. Mater.* **29** 1604764

Zhang J D, Sun X, Li H H, Ma H S, Duan F, Wu Z Y, Zhu B W, Chen R H and Nie L M 2023 In vivo characterization and analysis of glioblastoma at different stages using multiscale photoacoustic molecular imaging *Photoacoustics* **30** 100462

Zhang P, Li L, Lin L, Shi J and Wang L V 2019 In vivo superresolution photoacoustic computed tomography by localization of single dyed droplets *Light: Sci. Appl.* **8** 36

Zhang Q, Liu Z, Carney P R, Yuan Z, Chen H, Roper S N and Jiang H 2008 Non-invasive imaging of epileptic seizures *in vivo* using photoacoustic tomography *Phys. Med. Biol.* **53** 1921

Zhang S W *et al* 2022 The global burden and associated factors of ovarian cancer in 1990-2019: findings from the Global Burden of Disease Study 2019 *BMC Public Health* **22** 1455

Zhang T, Guo H, Qi W and Xi L 2024 Wearable photoacoustic watch for humans *Opt. Lett.* **49** 1524–7

Zhou Y, Zhang C, Yao D K and Wang L V 2012 Photoacoustic microscopy of bilirubin in tissue phantoms *J. Biomed. Opt.* **17** 126019

Zhu H, Isikman S O, Mudanyali O, Greenbaum A and Ozcan A 2013 Optical imaging techniques for point-of-care diagnostics *Lab Chip* **13** 51–67

IOP Publishing

Diagnostic Biomedical Optics
Fundamentals and applications
Murukeshan Vadakke Matham, C S Suchand Sandeep, Maria Merin Antony, Manojit Pramanik and Santhosh Chidangil

Chapter 7

Conclusion and reflection

Murukeshan Vadakke Matham

This chapter highlights the contributions and impact of various imaging modalities, each offering unique strengths and applications. With optical imaging technologies at the forefront, these methods harness light to visualize structures and processes within biological tissues, showcasing their versatility and transformative potential in diagnostics.

7.1 Conclusion

The field of diagnostic optics and medical imaging has advanced significantly, driven by the demand for precise, non-invasive diagnostic tools that offer detailed insights into biological structures, pathological and physiological conditions. This chapter summarizes the contributions and the impact of various imaging modalities, each with unique capabilities, strengths, and applications. Optical imaging technologies leverage light to visualize structures and processes in biological tissues. These techniques provide high spatial resolution and sensitivity, making them invaluable for observing cellular and molecular changes. Optical diagnostics, often used in conjunction with computational techniques, allow clinicians and researchers to interpret complex biological phenomena with high accuracy, aiding in early diagnosis and personalized treatment planning. In chapter 2, studies using optics and the use of properties and behaviour of light that include light–matter interaction were analysed in detail. It also covers light-based instruments with a wide range of applications including imaging and diagnostics. This chapter gave detailed descriptions on some of the basic optics and optical components, light–matter interaction, basics of imaging, and related parameters that are used for imaging applications. Different illumination beam profiles used for imaging purposes such as Gaussian beams and Bessel beams will also be explained along with some of the important aspects pertaining to medical imaging.

doi:10.1088/978-0-7503-2364-2ch7

Optical imaging, through methods such as confocal microscopy and optical coherence tomography (OCT), has become a foundational tool for capturing detailed images at cellular and sub-cellular levels. Its non-invasive nature, combined with high-resolution imaging capabilities, makes it especially useful in fields like ophthalmology and dermatology. Although light scattering in tissues limits its imaging depth, recent advances, such as adaptive optics and phase-sensitive techniques, continue to extend its applications. Multimodality imaging integrates two or more imaging techniques to leverage their complementary strengths, allowing for more comprehensive diagnostic data. By combining anatomical, functional, and molecular information, multimodality approaches provide a broader view of the underlying pathology. Techniques such as PET/CT, MRI/CT, and the integration of optical imaging with ultrasound have become instrumental in oncology and neurology, where multiple data layers are essential for precise diagnostics. In this context, chapter 3 discusses major diagnostic biomedical optical imaging modalities and the principles behind them. The chapter starts with a discussion of wide-field imaging systems followed by confocal imaging systems. It then introduces photoacoustic microscopy, OCT, and spectroscopic imaging systems. It concludes with a discussion on structured illumination microscopy and laser speckle imaging. The advantages, limitations, and biomedical applications of each of these imaging modalities are detailed.

Hyperspectral imaging (HSI) captures a wide spectrum of wavelengths for each pixel in an image, providing detailed spectral information that reveals biochemical and structural differences within tissues. HSI is particularly useful in detecting subtle changes in tissue composition and pathology that are not visible in standard imaging. This technology has shown potential in areas like oncology, dermatology, and wound healing, where early detection of pathological changes is critical. Chapter 4 provided an in-depth exploration of multispectral and HSI, focusing on the principles, instrumentation, and applications of these technologies in various diagnostic imaging domains. The chapter is divided into several sections, each addressing key aspects of multispectral and hyperspectral imaging systems. Instrumentation of the HSI system is discussed in detail, covering the essential steps for setting up the equipment. This includes the spectral calibration of the CCD camera, which ensures accurate wavelength measurements, and the spatial calibration, which focuses on precise spatial alignment of the imaging system. The chapter also addresses the processes involved in data acquisition and processing, which are crucial for extracting meaningful information from HSI data. A section on resolution explains the importance of achieving high spatial and spectral clarity in HSI, which is essential for accurate analysis.

Raman spectroscopy is a molecular imaging technique based on inelastic scattering of light, providing a 'fingerprint' of molecular composition. This method is sensitive to biochemical changes at the molecular level, making it a powerful tool for identifying pathological changes in tissues, often before morphological alterations become apparent. It has applications in cancer diagnostics, metabolic disorder studies, and drug monitoring, offering a non-invasive means to obtain real-time molecular data. From these perspectives, chapter 5 presented the most popular Raman spectroscopy techniques, such as micro-Raman spectroscopy, surface-enhanced Raman spectroscopy and Raman tweezers spectroscopy, and some of their medical applications. These techniques offer valuable insights into important biomolecular species in human clinical samples and their functions under different health conditions. The clinical samples selected for the study consist of blood components, cells and tissues. Raman-based techniques to be discussed here are non-destructive and require a minimal quantity of samples, and sample preparation is relatively easy, making it advantageous for various analytical applications.

Photoacoustic imaging (PAI) combines optical and ultrasound imaging principles to offer high-resolution, deep-tissue images. By using light to generate acoustic signals, this technique overcomes the depth limitations of traditional optical imaging while preserving spatial resolution. It is especially promising in oncology and vascular imaging, where visualization of blood vessel networks and oxygen saturation can provide crucial insights into disease progression and treatment efficacy. Chapter 6 introduced an emerging hybrid imaging modality known as PAI for pre-clinical and clinical applications. A brief description of and rationale for PAI will be given. The different types of clinical/pre-clinical PAI systems will be described along with their respective strengths and weaknesses. The types of clinical systems discussed are categorized into whether they perform tomography, microscopy, and endoscopy. Much of the chapter will focus on recent applications of PAI categorized by human organ from a head-to-toe fashion. We will conclude with a look at surgical and monitoring devices based on PAI and throw out some prospects of this new technology.

7.2 Future directions

Medical imaging techniques like MRI, CT, and ultrasound have revolutionized diagnostics by enabling visualization of internal structures non-invasively and in real time. These methods are fundamental in nearly every aspect of modern medicine, from screening and diagnosis to treatment planning and monitoring. Each modality has strengths—MRI's soft tissue contrast, CT's rapid imaging speed, and ultrasound's real-time capabilities that can be optimized based on clinical needs. While these techniques are well-established, ongoing research continues to refine their resolution, contrast, and speed.

The evolution of these imaging modalities points toward a future, where multimodal and molecular imaging are seamlessly integrated into clinical workflows. The combination of high-resolution anatomical imaging with molecular specificity could transform personalized medicine, enabling earlier and more precise diagnoses. Advances in artificial intelligence and machine learning further enhance the potential of these techniques by improving image interpretation, reducing operator dependency, and providing quantitative insights.

In general, the future of diagnostic biomedical optics is promising, with advances in imaging techniques, artificial intelligence, and sensor technology. Here's a look at some key trends and possibilities: Techniques like OCT and multiphoton microscopy are evolving to provide even more detailed imaging of tissue structures at cellular and molecular levels (Murukeshan *et al* 2015, Ratheesh *et al* 2016a, 2016b). This allows early detection of diseases like cancer, where cellular changes are critical markers (Dinish *et al* 2007, Pae *et al* 2021).

Super-resolution microscopy techniques (e.g., STED, SIM) will likely become more accessible, aiding in the visualization of fine cellular structures previously undetectable by traditional optics (Neil *et al* 1997, Perinchery *et al* 2019, Antony *et al* 2023a). AI and machine learning will enhance diagnostic imaging interpretation, identifying subtle patterns in optical data that might escape the human eye. For instance, in OCT or hyperspectral imaging, algorithms can detect changes associated with disease progression. Deep learning could also play a role in classifying tissue samples, assessing cell abnormalities, and providing predictive analytics for treatment outcomes (Lu and Fei 2014, Lim and Murukeshan 2016a, Ratheesh *et al* 2016a, Lim and Murukeshan 2017a, Suchand Sandeep *et al* 2019, Merin Antony *et al* 2020, Antony *et al* 2023b, 2024, Merin Antony *et al* 2024).

Advances in miniaturization are allowing for the development of portable, even wearable, diagnostic devices. Devices using optical sensors (e.g., near-infrared spectroscopy for glucose monitoring) may provide continuous, non-invasive diagnostics, particularly useful in managing chronic diseases. Portable imaging devices will enable bedside diagnostics and point-of-care testing, transforming patient care in both hospital and remote settings real time (Murukeshan *et al* 2015, Lim and Murukeshan 2017b).

Techniques combining multiple optical methods—like fluorescence, Raman spectroscopy, and infrared spectroscopy—offer comprehensive views of tissue characteristics, including metabolic, structural, and chemical information. This ‘multiparametric’ approach could allow clinicians to gather information on tissue health, oxygenation, and even molecular composition in real-time, aiding in more precise diagnoses.

Biomedical optics can play a crucial role in monitoring immune responses to treatments, especially immunotherapies. Using optical techniques, clinicians may detect early responses to therapy, track tumor microenvironment changes, and better personalize cancer treatments. Spectroscopic techniques, such as photoacoustic and Raman spectroscopy, are becoming capable of performing non-invasive ‘optical biopsies.’, as detailed in the previous chapters. These methods could replace traditional biopsy procedures in some cases, reducing patient discomfort and speeding up diagnosis.

Development of new contrast agents and molecular probes is expanding the capabilities of optical diagnostics. For example, nanoscale probes can target specific biomarkers, improving the sensitivity and specificity of diagnostics. Furthermore, smart probes that respond to environmental conditions (pH, temperature, etc) could help in the real-time monitoring of disease processes or therapeutic effects. As optical devices become more embedded in consumer electronics (like smartwatches), biomedical optics can leverage continuous health monitoring data. Combined with optical sensors, these devices could provide alerts about health issues in real time, promoting preventative care and early interventions (Murukeshan and Sujatha

2007, Murukeshan *et al* 2015, Lim and Murukeshan 2016a, 2016b, Merin Antony *et al* 2020).

A major challenge in optical imaging has been the inability to image around opaque obstacles. This limitation has significant implications for medical applications, such as injections, with an estimated 20 billion injections administered globally each year. Blind injections or improper needle positioning often led to needle-stick injuries due to the needle blocking the field of view during insertion. Similarly, surgical tools can obstruct the view, particularly when dealing with small anatomical structures (Ale *et al* 2012). Thus, there is a pressing need to develop imaging systems that can capture images around opaque obstacles, which could greatly enhance medical therapeutics and reduce procedural risks.

Overall, biomedical optics is on a path to becoming less invasive, more precise, and more integrated with other digital health technologies. This convergence diagnostic imaging and the above-mentioned disruptive innovations happening in this field will likely transform diagnostics by making it faster, more personalized, and more accessible, driving progress towards attaining patient-centred healthcare.

7.3 Summary

In conclusion, diagnostic optics and medical imaging encompass a range of technologies that offer unique insights into biological and pathological processes. By leveraging their strengths and addressing their limitations, these imaging modalities provide a roadmap for future innovations in medical diagnostics. Continued research, coupled with technological advancements, promises to expand the boundaries of what these modalities can achieve, paving the way for better diagnostic accuracy, earlier disease detection, and ultimately, improved patient outcomes.

References

Ale A, Ermolayev V, Herzog E, Cohrs C, de Angelis M H and Ntziachristos V 2012 FMT-XCT: in vivo animal studies with hybrid fluorescence molecular tomography-X-ray computed tomography *Nat. Methods* **9** 615–20

Antony M M, Haridas A, Suchand Sandeep C S and Vadakke Matham M 2023a An optodigital system for visualizing the leaf epidermal surface using embedded speckle SIM: a 3D non-destructive approach *Comput. Electron. Agric.* **211** 107962

Antony M M, Suchand Sandeep C S, Lim H-T and Vadakke Matham M 2023b High-resolution ultra-spectral imager for advanced imaging in agriculture and biomedical applications *J. Biomed. Photon. Eng.* **9** 8928

Antony M M, Suchand Sandeep C S and Vadakke Matham M 2024 Hyperspectral vision beyond 3D: a review *Opt. Lasers Eng.* **178** 108238

Dinish U S, Gulati P, Murukeshan V M and Seah L K 2007 Diagnosis of colon cancer using frequency domain fluorescence imaging technique *Opt. Commun.* **271** 291–301

Lim H-T and Murukeshan V M 2017a Hyperspectral imaging of polymer banknotes for building and analysis of spectral library *Opt. Lasers Eng.* **98** 168–75

Lim H-T and Murukeshan V M 2017b Hyperspectral photoacoustic spectroscopy of highly-absorbing samples for diagnostic ocular imaging applications *Int. J. Optomechatronics* **11** 36–46

Lim H T and Murukeshan V M 2016a A four-dimensional snapshot hyperspectral video-endoscope for bio-imaging applications *Sci. Rep.* **6** 24044

Lim H T and Murukeshan V M 2016b Spatial-scanning hyperspectral imaging probe for bio-imaging applications *Rev. Sci. Instrum.* **87** 033707

Lu G and Fei B 2014 Medical hyperspectral imaging: a review *J. Biomed. Opt.* **19** 10901

Merin Antony M, Suchand Sandeep C S and Matham M V 2020 Probe-based hyperspectral imager for crop monitoring *Proc. SPIE 11525, SPIE Future Sensing Technologies* **1152512**

Merin Antony M, Suchand Sandeep C S, Bijeesh M M and Matham M V 2024 A fast analysis approach for crop health monitoring in hydroponic farms using hyperspectral imaging *Proc. SPIE 12879, Photonic Technologies in Plant and Agricultural Science* **12879** 128790G

Murukeshan V M, Jesmond H X J, Shinoj V K, Baskaran M and Tin A 2015 Non-contact high resolution Bessel beam probe for diagnostic imaging of cornea and trabecular meshwork region in eye *Clinical and Biomedical Spectroscopy and Imaging IV* ed J Brown and V Deckert *Proc. SPIE* **9537** 953728

Murukeshan V M and Sujatha N 2007 All fiber based multispeckle modality endoscopic system for imaging medical cavities *Rev. Sci. Instrum.* **78** 053106

Neil M A, Juskaitis R and Wilson T 1997 Method of obtaining optical sectioning by using structured light in a conventional microscope *Opt. Lett.* **22** 1905–7

Pae J Y, Nair R V, Padmanabhan P, Radhakrishnan G, Gulyas B and Vadakke Matham M 2021 Gold nano-urchins enhanced surface plasmon resonance (SPR) biosensors for the detection of estrogen receptor alpha (ERα) *IEEE J. Sel. Top. Quantum Electron.* **27** 1–6

Perinchery S M, Haridas A, Shinde A, Buchnev O and Murukeshan V M 2019 Breaking diffraction limit of far-field imaging via structured illumination Bessel beam microscope (SIBM) *Opt. Express* **27** 6068–82

Ratheesh K M, Prabhathan P, Seah L K and Murukeshan V M 2016a Gold nanorods with higher aspect ratio as potential contrast agent in optical coherence tomography and for photo-thermal applications around 1300 nm imaging window *Biomed. Phys. Eng. Express* **2** 055005

Ratheesh K M, Seah L K and Murukeshan V M 2016b Spectral phase-based automatic calibration scheme for swept source-based optical coherence tomography systems *Phys. Med. Biol.* **61** 7652–63

Suchand Sandeep C S, Sarangapani S, Hong X J J, Aung T, Baskaran M and Murukeshan V M 2019 Optical sectioning and high resolution visualization of trabecular meshwork using Bessel beam assisted light sheet fluorescence microscopy *J. Biophotonics* **12** e201900048

www.ingramcontent.com/pod-product-compliance
Ingram Content Group UK Ltd.
Pitfield, Milton Keynes, MK11 3LW, UK
UKHW051852100926
8556IPUK00005B/585

9 780750 3